Write your Username ✏

Password ✏

Get your resources online to help you in class and at home.

Vocab

Find activities for building vocabulary.

Watch

Watch animations of key concepts.

Tools

Explore concepts with virtual manipulatives.

Check

Self-assess your progress.

eHelp

Get targeted homework help.

Games

Reinforce with games and apps.

Tutor

See a teacher illustrate examples and problems.

GO mobile

Scan this QR code with your smart phone* or visit mheonline.com/stem_apps.

*May require quick response code reader app.

v

Contents in Brief
Organized by Domain

Common Core State Standards

Operations and Algebraic Thinking

Chapter 1   Apply Addition and Subtraction Concepts
Chapter 2   Number Patterns
Chapter 3   Add Two-Digit Numbers
Chapter 4   Subtract Two-Digit Numbers

Number and Operations in Base Ten

Chapter 5   Place Value to 1,000
Chapter 6   Add Three-Digit Numbers
Chapter 7   Subtract Three-Digit Numbers

Measurement and Data

Chapter 8   Money
Chapter 9   Data Analysis
Chapter 10   Time
Chapter 11   Customary and Metric Lengths

Geometry

Chapter 12   Geometric Shapes and Equal Shares

Standards for Mathematical PRACTICE Woven Throughout

Chapter 1

Apply Addition and Subtraction Concepts

ESSENTIAL QUESTION
What strategies can I use to add and subtract?

Getting Started

Am I Ready? 3
My Math Words 4
My Vocabulary Cards 5
My Foldable **FOLDABLES** 9

I ♥ animals!

Lessons and Homework

Lesson 1 Addition Properties 11
Lesson 2 Count On to Add 17
Lesson 3 Doubles and Near Doubles 23
Lesson 4 Make a 10 29
Lesson 5 Add Three Numbers 35
Lesson 6 Problem-Solving Strategy:
Write a Number Sentence 41
Check My Progress 47
Lesson 7 Count Back to Subtract 49
Lesson 8 Subtract All and Subtract Zero 55
Lesson 9 Use Doubles to Subtract 61
Check My Progress 67
Lesson 10 Relate Addition and Subtraction 69
Lesson 11 Missing Addends 75
Lesson 12 Fact Families 81
Lesson 13 Two-Step Word Problems 87

Wrap Up

Fluency Practice 93
My Review 95
Reflect 98

connectED.mcgraw-hill.com

Watch ▶
Look for this!
Click online and you can watch videos that will help you learn the lessons.

Chapter 2 Number Patterns

ESSENTIAL QUESTION
How can equal groups help me add?

Getting Started

Am I Ready? 101
My Math Words 102
My Vocabulary Cards 103
My Foldable **FOLDABLES** 105

Lessons and Homework

Lesson 1   Skip Count on a Hundred Chart 107
Lesson 2   Skip Count by 2s, 5s, and 10s 113
Lesson 3   Problem-Solving Strategy: Find a Pattern 119
Check My Progress 125
Lesson 4   Repeated Addition 127
Lesson 5   Repeated Addition with Arrays 133
Lesson 6   Even and Odd Numbers 139
Lesson 7   Sums of Equal Numbers 145

Wrap Up

Fluency Practice 151
My Review 153
Reflect 156

Welcome to the desert, cowpoke!

Hi!

connectED.mcgraw-hill.com

Chapter 3
Add Two-Digit Numbers

Copyright © The McGraw-Hill Companies, Inc. (tr)Jules Frazier/Photodisc/Getty Images, (bl)Ocean/CCRBIS, (br)Prestock/SuperFusion/SuperStock

ESSENTIAL QUESTION
How can I add
two-digit numbers?

Getting Started

Am I Ready? 159
My Math Words 160
My Vocabulary Cards 161
My Foldable **FOLDABLES** 163

Lessons and Homework

Lesson 1   Take Apart Tens to Add 165
Lesson 2   Regroup Ones as Tens 171
Lesson 3   Add to a Two-Digit Number 177
Check My Progress 183
Lesson 4   Add Two-Digit Numbers 185
Lesson 5   Rewrite Two-Digit Addition 191
Lesson 6   Add Three or Four Two-Digit Numbers 197
Lesson 7   Problem-Solving Strategy: Make a Model 203

Wrap Up

Fluency Practice 209
My Review 211
Reflect 214

We make a
great pair!

connectED.mcgraw-hill.com

eHelp **Look for this!**
Click online and you
can get more help
while doing your
homework.

Chapter 4 Subtract Two-Digit Numbers

ESSENTIAL QUESTION
How can I subtract two-digit numbers?

Getting Started

Am I Ready? 217
My Math Words 218
My Vocabulary Cards 219
My Foldable FOLDABLES 221

Lessons and Homework

Lesson 1   Two-Digit Fact Families 223
Lesson 2   Take Apart Tens to Subtract 229
Lesson 3   Regroup a Ten as Ones 235
Lesson 4   Subtract From a Two-Digit Number 241
Check My Progress 247
Lesson 5   Subtract Two-Digit Numbers 249
Lesson 6   Rewrite Two-Digit Subtraction 255
Lesson 7   Check Subtraction 261
Lesson 8   Problem-Solving Strategy:
           Write a Number Sentence 267
Lesson 9   Two-Step Word Problems 273

Wrap Up

Fluency Practice 279
My Review 281
Reflect 284

Springtime!

connectED.mcgraw-hill.com

Chapter 5 — Place Value to 1,000

ESSENTIAL QUESTION
How can I use place value?

Getting Started

Am I Ready? 287
My Math Words 288
My Vocabulary Cards 289
My Foldable FOLDABLES 293

Lessons and Homework

Lesson 1   Hundreds 295
Lesson 2   Hundreds, Tens, and Ones 301
Lesson 3   Place Value to 1,000 307
Check My Progress 313
Lesson 4   Problem-Solving Strategy:
            Use Logical Reasoning 315
Lesson 5   Read and Write Numbers to 1,000 321
Lesson 6   Count by 5s, 10s, and 100s 327
Lesson 7   Compare Numbers to 1,000 333

Wrap Up

My Review 339
Reflect 342

Look how I've grown!

connectED.mcgraw-hill.com

Look for this!
Click online and you can find tools that will help you explore concepts.

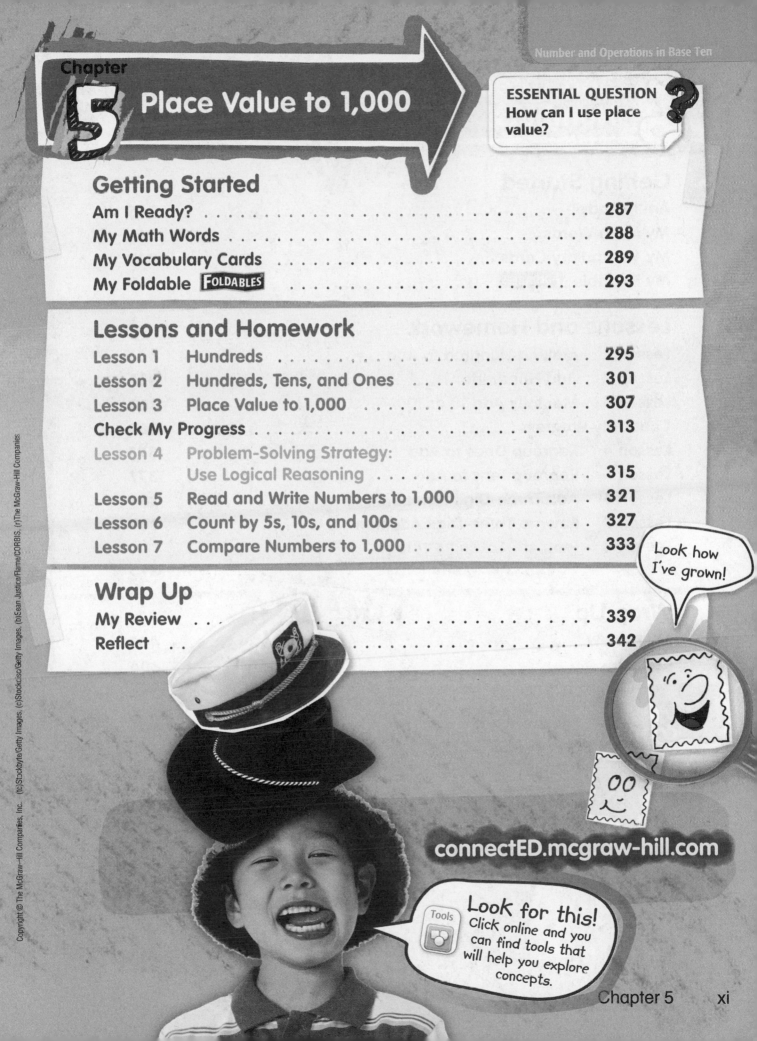

Chapter 6 Add Three-Digit Numbers

ESSENTIAL QUESTION
How can I add three-digit numbers?

Getting Started

Am I Ready? 345
My Math Words 346
My Vocabulary Cards 347
My Foldable FOLDABLES 349

Lessons and Homework

Lesson 1   Make a Hundred to Add 351
Lesson 2   Add Hundreds 357
Lesson 3   Mentally Add 10 or 100 363
Check My Progress 369
Lesson 4   Regroup Ones to Add 371
Lesson 5   Regroup Tens to Add 377
Lesson 6   Add Three-Digit Numbers 383
Lesson 7   Rewrite Three-Digit Addition 389
Lesson 8   Problem-Solving Strategy:
            Guess, Check, and Revise 395

Wrap Up

My Review 401
Reflect 404

MAIN ST.

LITTLE D.

connectED.mcgraw-hill.com

My house

Chapter 7 Subtract Three-Digit Numbers

ESSENTIAL QUESTION
How can I subtract three-digit numbers?

Getting Started

Am I Ready? 407
My Math Words 408
My Vocabulary Cards 409
My Foldable FOLDABLES 411

Lessons and Homework

Lesson 1   Take Apart Hundreds to Subtract 413
Lesson 2   Subtract Hundreds 419
Lesson 3   Mentally Subtract 10 or 100 425
Check My Progress 431
Lesson 4   Regroup Tens 433
Lesson 5   Regroup Hundreds 439
Lesson 6   Subtract Three-Digit Numbers 445
Lesson 7   Rewrite Three-Digit Subtraction 452
Lesson 8   Problem-Solving Strategy:
            Write a Number Sentence 457
Lesson 9   Subtract Across Zeros 463

Wrap Up

My Review 469
Reflect 472

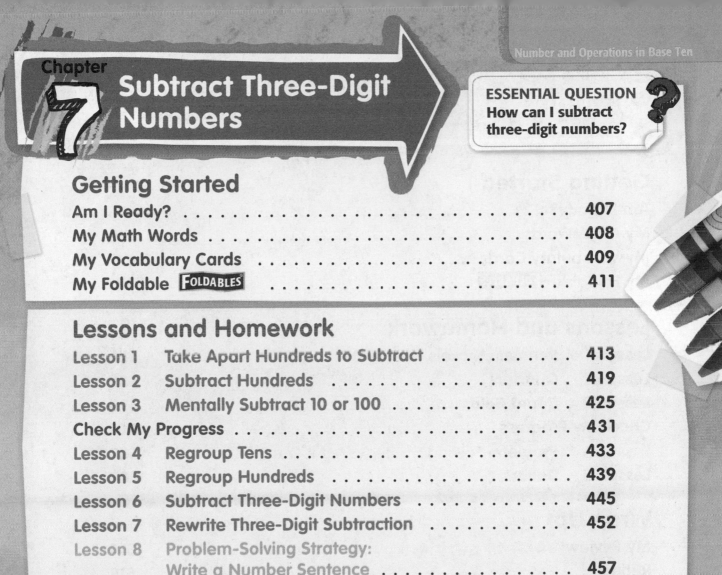

Vocab
Look for this!
Click online and you can find activities to help build your vocabulary.

You + School = Cool

connectED.mcgraw-hill.com

Chapter

8 Money

ESSENTIAL QUESTION
How do I count and use money?

Getting Started

Am I Ready? **475**
My Math Words **476**
My Vocabulary Cards **477**
My Foldable FOLDABLES **481**

Lessons and Homework

Lesson 1   Pennies, Nickels, and Dimes **483**
Lesson 2   Quarters **489**
Lesson 3   Count Coins **495**
Check My Progress **501**
Lesson 4   Problem-Solving Strategy: Act it Out **503**
Lesson 5   Dollars **509**

Wrap Up

My Review **515**
Reflect **518**

You make me laugh!

Look for this!
Check ✓ Click online and you can check your progress.

connectED.mcgraw-hill.com

Copyright © The McGraw-Hill Companies, Inc. (l)Creative Crop/Digital Vison/Getty Images, (r)Thinkstock/Comstock/Getty Images

Chapter 9 Data Analysis

ESSENTIAL QUESTION
How can I record and analyze data?

Getting Started

Am I Ready? 521
My Math Words 522
My Vocabulary Cards 523
My Foldable FOLDABLES 527

Go nuts!

Lessons and Homework

Lesson 1   Take a Survey 529
Lesson 2   Make Picture Graphs 535
Lesson 3   Analyze Picture Graphs 541
Check My Progress 547
Lesson 4   Make Bar Graphs 549
Lesson 5   Analyze Bar Graphs 555
Lesson 6   Problem-Solving Strategy: Make a Table ... 561
Lesson 7   Make Line Plots 567
Lesson 8   Analyze Line Plots 573

Wrap Up

My Review 579
Reflect 582

connectED.mcgraw-hill.com

Chapter 10 Time

ESSENTIAL QUESTION
How do I use and tell time?

Getting Started

Am I Ready? 585
My Math Words 586
My Vocabulary Cards 587
My Foldable FOLDABLES 591

Lessons and Homework

Lesson 1 Time to the Hour 593
Lesson 2 Time to the Half Hour 599
Lesson 3 Problem-Solving Strategy: Find a Pattern 605
Check My Progress 611
Lesson 4 Time to the Quarter Hour 613
Lesson 5 Time to Five Minute Intervals 619
Lesson 6 A.M. and P.M. 625

Wrap Up

My Review 631
Reflect 634

It's time for our trip!

connectED.mcgraw-hill.com

Chapter 11 Customary and Metric Lengths

I love rulers because I'm an inch worm!

Getting Started

Am I Ready? 637
My Math Words 638
My Vocabulary Cards 639
My Foldable FOLDABLES 643

Lessons and Homework

Lesson 1   Inches 645
Lesson 2   Feet and Yards 651
Lesson 3   Select and Use Customary Tools 657
Check My Progress 663
Lesson 4   Compare Customary Lengths 665
Lesson 5   Relate Inches, Feet, and Yards 671
Lesson 6   Problem-Solving Strategy:
            Use Logical Reasoning 677
Check My Progress 683
Lesson 7   Centimeters and Meters 685
Lesson 8   Select and Use Metric Tools 691
Lesson 9   Compare Metric Lengths 697
Lesson 10   Relate Centimeters and Meters 703
Lesson 11   Measure on a Number Line 709
Lesson 12   Measurement Data 715

Wrap Up

My Review 721
Reflect 724

connectED.mcgraw-hill.com

Chapter 12
Geometric Shapes and Equal Shares

ESSENTIAL QUESTION
How do I use shapes and equal parts?

Getting Started

Am I Ready? 727
My Math Words 728
My Vocabulary Cards 729
My Foldable FOLDABLES 737

Lessons and Homework

Lesson 1   Two-Dimensional Shapes 739
Lesson 2   Sides and Angles 745
Lesson 3   Problem-Solving Strategy: Draw a Diagram 751
Check My Progress 757
Lesson 4   Three-Dimensional Shapes 759
Lesson 5   Faces, Edges, and Vertices 765
Lesson 6   Relate Shapes and Solids 771
Lesson 7   Halves, Thirds, and Fourths 777
Lesson 8   Area 783

Wrap Up

My Review 789
Reflect 792

Come on guys. Let's check it out!

connectED.mcgraw-hill.com

Chapter

8 Money

Let's Count Money!

Watch a video!

My Common Core State Standards

CCSS

2.MD.8 Solve word problems involving dollar bills, quarters, dimes, nickels, and pennies, using $ and ¢ symbols appropriately.

Standards for Mathematical PRACTICE ⬇

1. Make sense of problems and persevere in solving them.
2. Reason abstractly and quantitatively.
3. Construct viable arguments and critique the reasoning of others.
4. Model with mathematics.
5. Use appropriate tools strategically.
6. Attend to precision.
7. Look for and make use of structure.
8. Look for and express regularity in repeated reasoning.

= focused on in this chapter

Name _____

 ← Go online to take the Readiness Quiz

1. Skip count by 5s.

5, 10, __15__ , __20__ , __25__ , __30__

2. Skip count by 10s.

10, 20, __30__ , __40__ , __50__ , __60__

Count on to add.

3. 25
   + 1
   ‾‾‾‾
    26

4. 30
   + 2
   ‾‾‾‾
    32

5. 18
   + 3
   ‾‾‾‾
    21

6. 16
   + 4
   ‾‾‾‾
    20

7. 20
   + 5
   ‾‾‾‾
    25

8. 35
   +10
   ‾‾‾‾
    45

9. Kim is counting her mittens. She skip counts by 2s. She counts up to 12. How many pairs of mittens does Kim have?

__6__ pairs

How Did I Do? Shade the boxes to show the problems you answered correctly.

1 2 3 4 5 6 7 8 9

My Math Words

Vocab
abc

Review Vocabulary

equal groups   repeated addition   skip count

Look at each example. Complete each sentence to describe the example. Then write the repeated addition sentence.

Example	Describe	Write

Equal groups of ___4___.

Skip count by ___2___.

2+2+2+2

Equal groups of ___5___.

Skip count by ___5___.

5+5+5+5+5

Equal groups of ___2___.

Skip count by ___10___.

10 +10

My Vocabulary Cards

Lesson 8–1

cent

→ 1 cent

→ 5 cents

Lesson 8–1

cent sign (¢)

1 ¢   5 ¢

Lesson 8–3

dime

10 cents
10¢

Lesson 8–7

dollar

100 cents

100 ¢

Lesson 8–7

dollar sign ($)

$1 or $1.00

Lesson 8–2

nickel

5 cents
5¢

Directions:
Ideas for Use

- Have students think of words that rhyme with some of the words.

- Arrange the cards in alphabetical order.

- Tell students to create riddles for each word. Ask them to work with a friend to guess the word for each word card.

The sign used to show cents.

One penny equals one cent or 1¢. 5 pennies equals 5 cents or 5¢.

One dollar has a value of 100 cents or 100¢.

A dime has a value of 10 cents or 10¢.

A nickel has a value of five cents or 5¢.

The sign used to show dollars.

My Vocabulary Cards

Vocab
abc

Mathematical
PRACTICE

Lesson 0–1

penny

1 cent
1¢

Lesson 8–4

quarter

25 cents
25¢

Directions:
Ideas for Use

• Have students use a blank card to write this chapter's essential question. Have them use the back of the card to write or draw examples that help them answer the question.

• Use the blank cards to write your own vocabulary cards.

A quarter has a value of 25 cents or 25¢.

A penny has a value of one cent or 1¢.

Mia

Jasmine

Ella

Ryan

Nathan

Robert

Name

Pennies, Nickels, and Dimes

Lesson 1

ESSENTIAL QUESTION
How do I count and use money?

Explore and Explain

Watch   Tools

 Teacher Directions: Use pennies, nickels, and dimes. Sort the coins. Find the value of each group of coins. Write the value on each bouncy ball machine.

Online Content at connectED.mcgraw-hill.com

See and Show

Helpful Hint
¢ stands for cents.

Mathematical
PRACTICE

dime = 10¢

Count by 10s.

__10__¢ __20__¢

nickel = 5¢

Count by 5s.

__5__¢ __10__¢

penny = 1¢

Count by 1s.

__1__¢ __2__¢

To find the value of coins, start counting with the coin that has the greatest value.

__10__¢, __20__¢, __25__¢, __30__¢, __31__¢, __32__¢ = __32__¢

Count to find the value of the coins.

1.

__10__¢, __5__¢, __1__¢   = __16__¢

2.

__10__¢, __10__¢, __10__¢, __5__¢, __5__¢, __1__¢ = __41__¢

Talk Math   How many dimes are equal to 70 cents?

Name Amelia

On My Own

Count to find the value of the coins.

I'm having a ball counting coins!

3.

____10____ ¢,   ____10____ ¢,   ____5____ ¢,   ____5____ ¢,   ____1____ ¢   = ____31____ ¢

4.

____10____ ¢,   ____10____ ¢,   ____10____ ¢,   ____5____ ¢,   ____5____ ¢,   ____5____ ¢,   ____1____ ¢   = ____26____ ¢

5.

____10____ ¢,   ____5____ ¢,   ____5____ ¢,   ____1____ ¢,   ____1____ ¢,   ____1____ ¢,   ____1____ ¢   = ____24____ ¢

6.

____10____ ¢,   ____10____ ¢,   ____5____ ¢,   ____5____ ¢,   ____1____ ¢,   ____1____ ¢,   ____1____ ¢   = ____33____ ¢

Problem Solving

7. Jen had 6 dimes and 4 nickels. She lost 2 of each of them. How much does she still have?

$6 - 2 = 4$

_~~86~~_ ¢

8. Marcy wants to buy beads that cost 80¢ to make a friendship bracelet. If she has 2 nickels, how many dimes does she need to buy the beads?

_7_ dimes

9. Derek has some dimes. He gives Luis 4 dimes. He gives Mia 3 dimes. How much money did Derek give away?

_7_ ¢

HOT Problem Paul finds 5 dimes and 2 nickels. He counts them and says he has 50¢. Tell why Paul is wrong. Make it right.

because he added $3 + 2 = 50$ insted of $5 + 2 = 70$ ₵.

Name _____

My Homework

Homework Helper

eHelp

Need help? connectED.mcgraw-hill.com

dime = 10¢

Count by 10s.

10¢,   20¢

nickel = 5¢

Count by 5s.

5¢,   10¢

penny = 1¢

Count by 1s.

1¢,   2¢

Helpful Hint
¢ stands for cents.

To find the value of coins, start counting with the coin that has the greatest value.

10¢,   20¢,   25¢,   30¢,   31¢,   32¢ = 32¢

Count to find the value of the coins.

1.

__10__ ¢,   __10__ ¢,   __10__ ¢,   __5__ ¢,   __1__ ¢ = __36__ ¢

2.

__10__ ¢,   __5__ ¢,   __5__ ¢,   __5__ ¢,   __1__ ¢,   __1__ ¢ = __27__ ¢

Count to find the value of the coins.

We're TOPS at finding coin values!

3.

___10___ ¢, ___10___ ¢, ___10___ ¢, ___5___ ¢, ___1___ ¢, ___1___ ¢ = ___37___ ¢

4.

___10___ ¢, ___5___ ¢, ___5___ ¢, ___5___ ¢, ___5___ ¢, ___1___ ¢ = ___31___ ¢

5. Ken has 80¢. His friend has 4 dimes. How many nickels does his friend need to have the same amount of money as Ken?

___8___ nickels

Vocabulary Check

Circle the correct answer.

6. **dime**

Name

Quarters

Copyright © The McGraw-Hill Companies, Inc.   United States coin images from the United States Mint

Lesson 2

ESSENTIAL QUESTION
How do I count and use money?

Explore and Explain

Tools

$

MAGIC MONEY COUNTER

quarters
25¢

dimes
10¢

nickels
5¢

pennies
1¢

Teacher Directions: Use quarters, dimes, nickels, and pennies. Sort the coins into the correct columns. Count to find the value of the coins. Write the value on each column.

See and Show

quarter = 25¢

Count by 25s.

__25__¢, __50__¢, __75__¢

Start counting with the coin that has the greatest value.

_____¢, _____¢, _____¢, _____¢, _____¢, _____¢, _____¢ = _____¢

Count to find the value of the coins.

1. _____¢, _____¢, _____¢, _____¢, _____¢, _____¢ = _____¢

2. _____¢, _____¢, _____¢, _____¢, _____¢, _____¢ = _____¢

Talk Math How many quarters do you need to make 100¢?

Name _____

On My Own

Count to find the value of the coins.

3.

_____¢, _____¢, _____¢, _____¢, _____¢, _____¢, _____¢ = _____¢

4.

_____¢, _____¢, _____¢, _____¢ = _____¢

5.

_____¢, _____¢, _____¢, _____¢, _____¢, _____¢, _____¢ = _____¢

How many quarters do you need to purchase each item?

6.

50¢

_____ quarters

7.

PLAY TIME

100¢

_____ quarters

8.

75¢

_____ quarters

Problem Solving

Use the information to answer each question.

9. Dale found a quarter, a dime, and three nickels under the sofa. His mom gave him another quarter. Does he have enough money to buy a school basketball game ticket that costs 50¢?

_____

Can Dale also buy a juice box for 25¢?

_____

10. Lindsay has 2 quarters and 5 dimes. She gives her friend 1 quarter. Lindsay needs 100¢ to buy a stuffed animal. Does she have enough to buy the toy?

_____

11. Jan has 100¢ in quarters. She wants to buy bracelets. Each bracelet costs a quarter. How many bracelets can she buy?

_____

HOT Problem   Bryan buys water for 75¢. He uses 3 quarters. Describe another way Bryan could have paid for the water.

_____

_____

_____

My Homework

Lesson 2
Quarters

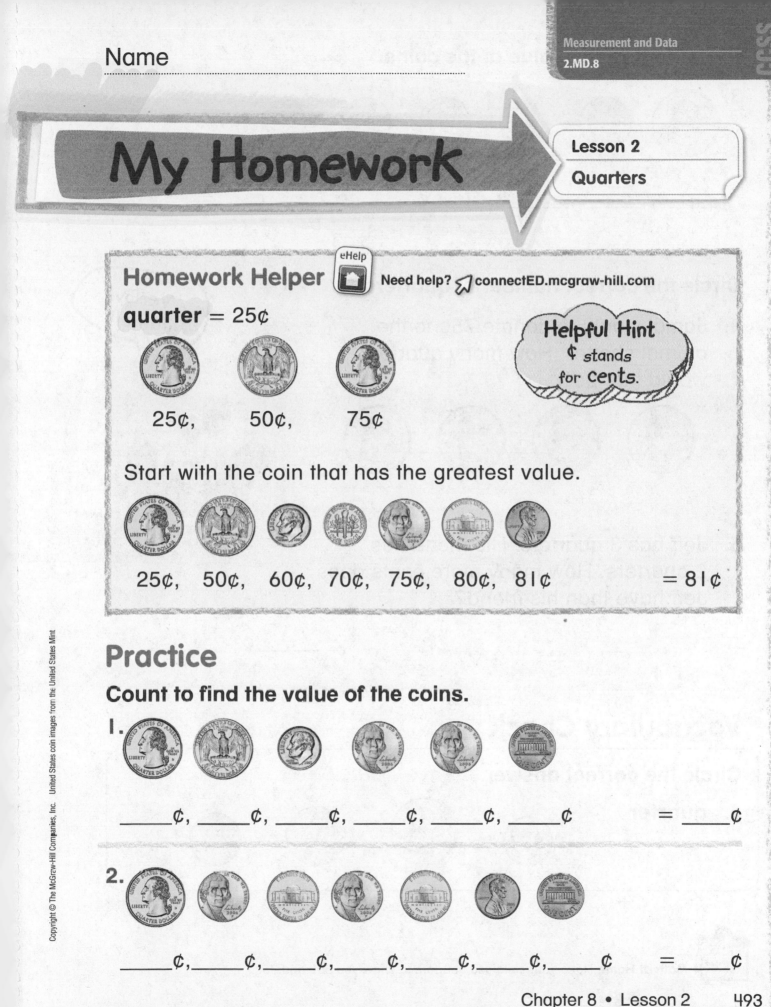

Homework Helper eHelp Need help? connectED.mcgraw-hill.com

quarter = 25¢

Helpful Hint
¢ stands for cents.

25¢,   50¢,   75¢

Start with the coin that has the greatest value.

25¢, 50¢, 60¢, 70¢, 75¢, 80¢, 81¢   = 81¢

Practice

Count to find the value of the coins.

1. _____¢, _____¢, _____¢, _____¢, _____¢, _____¢   = _____¢

2. _____¢, _____¢, _____¢, _____¢, _____¢, _____¢, _____¢   = _____¢

Count to find the value of the coins.

3.

_____¢, _____¢, _____¢, _____¢, _____¢, _____¢ = _____¢

Circle the correct number of quarters.

4. Jamal wants to donate 75¢ to the animal shelter. How many quarters would that be?

We love Jamal!

---

5. Jeff has 3 quarters. His friend has 2 quarters. How many more cents does Jeff have than his friend?

_____¢

Vocabulary Check

Vocab

Circle the correct answer.

6. **quarter**

Math at Home Have your child use quarters to show you 50¢ and 75¢.

Name _____

Count Coins

Lesson 3

ESSENTIAL QUESTION
How do I count and
use money?

Explore and Explain Tools

Quarters
25¢

Dimes
10¢

Nickels
5¢

Pennies
1¢

The value of all of the coins is _____.

Teacher Directions: Use quarters, dimes, nickels, and pennies. Sort the coins
into the appropriate columns. Trace them. Write the total value of the coins.

See and Show

Mathematical PRACTICE *Skip count!*

To count a group of coins, start with the coin that has the greatest value. Count to find the total.

(25¢) (25¢) (10¢) (1¢) (1¢)

25 ¢, **50** ¢, **60** ¢, **61** ¢, **62** ¢

= **62** ¢

Count to find the value of the coins.

1.

(10¢) (10¢) (5¢) (5¢) (5¢) (1¢)

_____¢,_____¢,_____¢,_____¢,_____¢,_____¢

= _____¢

2.

(25¢) (10¢) (10¢) (10¢) (5¢) (1¢)

_____¢,_____¢,_____¢,_____¢,_____¢,_____¢

= _____¢

Talk Math How does skip counting help you count groups of different coins?

Name

On My Own

Count to find the value of the coins.

3.

_____ ¢, _____ ¢, _____ ¢, _____ ¢, _____ ¢, _____ ¢

= _____ ¢

4.

_____ ¢, _____ ¢, _____ ¢, _____ ¢, _____ ¢, _____ ¢

= _____ ¢

Draw and label the coins from greatest to least.
Find the value of the coins.

5.

= _____ ¢

Problem Solving

6. Suppose you have 1 quarter, 3 dimes, 1 nickel, and 7 pennies. How much money do you have?

_____¢

7. Luke wants to buy a bouncy ball that costs 25 cents. He has five pennies, 1 dime, and 2 nickels. Does Luke have enough money?

_____

I'm earning it!

8. Connor has a quarter and a nickel. He gets 2 more quarters for helping around the house. How much money does he have now?

_____¢

 Chase has 5 dimes. Dan has 10 nickels. Who has more money? Explain.

_____

_____

_____

Name

My Homework

Lesson 3

Count Coins

Homework Helper

Need help? connectED.mcgraw-hill.com

To count coins, start with the coin that has the greatest value. Count to find the total value.

| 25¢ | 25¢ | 10¢ | 5¢ | 5¢ | 1¢ |

25¢,   50¢,   60¢,   65¢,   70¢,   71¢

= 71¢

Count to find the value of the coins.

1.

| 25¢ | 10¢ | 10¢ | 5¢ | 5¢ | 1¢ |

_____¢, _____¢, _____¢, _____¢, _____¢, _____¢

= _____¢

2.

| 25¢ | 25¢ | 10¢ | 10¢ | 5¢ | 5¢ | 5¢ |

_____¢, _____¢, _____¢, _____¢, _____¢, _____¢, _____¢

= _____¢

Count to find the value of the coins.

3.

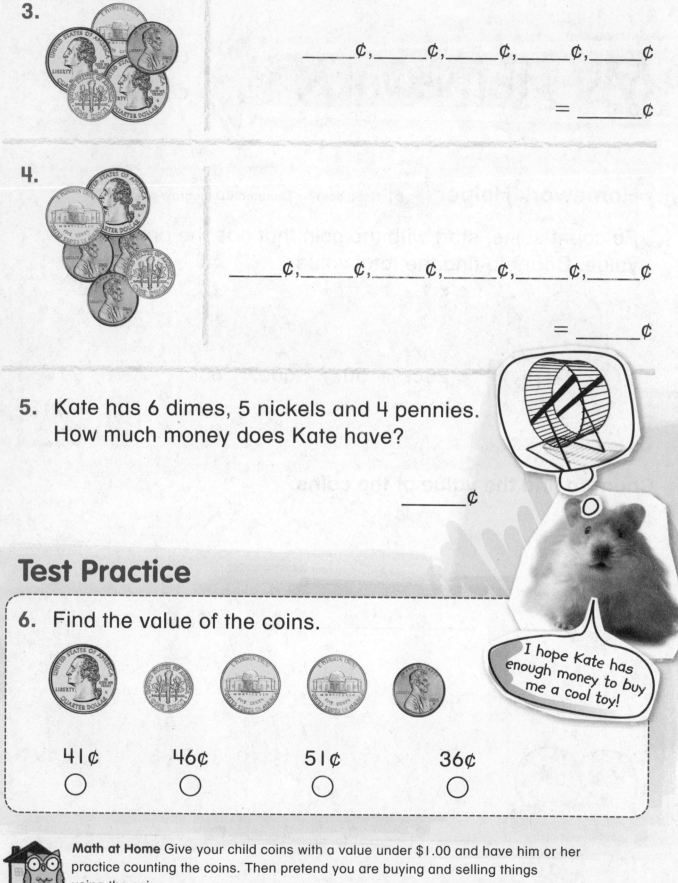

_____ ¢, _____ ¢, _____ ¢, _____ ¢, _____ ¢

= _____ ¢

4.

_____ ¢, _____ ¢, _____ ¢, _____ ¢, _____ ¢, _____ ¢

= _____ ¢

5. Kate has 6 dimes, 5 nickels and 4 pennies. How much money does Kate have?

_____ ¢

I hope Kate has enough money to buy me a cool toy!

Test Practice

6. Find the value of the coins.

41¢   46¢   51¢   36¢
○   ○   ○   ○

Math at Home Give your child coins with a value under $1.00 and have him or her practice counting the coins. Then pretend you are buying and selling things using the coins.

Check My Progress

Vocabulary Check

| penny | nickel | dime | quarter |

Complete each sentence.

1. A coin that has a value of 25 cents is a _____.

2. A coin that has a value of 5 cents is a _____.

3. A coin that has a value of 1 cent is a _____.

4. A coin that has a value of 10 cents is a _____.

Concept Check

Count to find the value of the coins.

5.

_____¢, _____¢, _____¢, _____¢, _____¢, _____¢, _____¢ = _____¢

6.

_____¢, _____¢, _____¢, _____¢, _____¢, _____¢, _____¢ = _____¢

Count to find the value of the coins.

7.

_____¢, _____¢, _____¢, _____¢, _____¢, _____¢   = _____¢

8.

_____¢, _____¢, _____¢, _____¢, _____¢, _____¢, _____¢, _____¢

= _____¢

9.

25¢   25¢   25¢   10¢   1¢   1¢

_____¢, _____¢, _____¢, _____¢, _____¢, _____¢

= _____¢

Test Practice

10. Myla needs 55¢ to buy a bag of popcorn.
Which coins should she use?

quarter, quarter
○

quarter, penny, nickel
○

quarter, quarter, nickel
○

quarter, quarter, penny
○

Problem Solving

STRATEGY: Act it Out

Watch | Tools

Gavin has 2 quarters, 1 dime, and 1 nickel. Does he have enough money to buy this toy?

Yippee, Gavin wants me!

65¢

1 Understand
Underline what you know.
Circle what you need to find.

2 Plan
How will I solve the problem?

3 Solve
Act it out.

(25¢)   (25¢)   (10¢)   (5¢)

25¢,   50¢,   60¢,   65¢   = ___65___ ¢

Does Gavin have enough money to buy the toy?

___yes___

4 Check
Is my answer reasonable? Explain.

Practice the Strategy

Robin wants to buy 3 rings.
Each ring costs 20¢. She has
1 quarter, 2 dimes, and 5 nickels.
Does she have enough money?

I need more fingers!

1 Understand
Underline what you know.
Circle what you need to find.

2 Plan
How will I solve the problem?

3 Solve
I will...

4 Check
Is my answer reasonable? Explain.

Name _____

Apply the Strategy

Your quarter's safe with me!

1. Maria has 1 quarter in her piggy bank. Her mom gives her a nickel. Her dad gives her a dime. How much money does Maria have in all?

_____

2. Mark has 2 quarters, 1 dime, and 1 penny. He wants to buy a toy truck for 55¢. Does he have enough money to buy the toy truck?

_____

3. Wesley has 2 quarters, 3 dimes, and 2 nickels. He has enough money to buy a race car. What is the greatest amount of money that the race car could cost?

_____

Review the Strategies

Choose a strategy
- Act it out.
- Draw a picture.
- Use logical reasoning.

4. Riley has 1 dime, 3 nickels, and 4 pennies. Does she have enough to buy a cookie that costs 30¢?

_____

How much more does she need?

_____

5. Annie has coins to buy a gel pen at the store. It costs 85¢. She has 2 quarters and 1 nickel. What two coins does she still need?

gel pen
85¢

_____

6. A notebook costs 40¢. What three coins could you use to pay for the toy?

_____

Name

My Homework

eHelp

Landon has 3 dimes and 2 nickels. Does he have enough money to buy the pack of stickers that costs 50¢?

50¢

1 Understand
Underline what you know.
Circle what you need to find.

2 Plan
How will I solve the problem?

3 Solve
Act it out.

10¢,   20¢,   30¢,   35¢,   40¢ = 40¢

He has 40¢. He needs 50¢.
So, he does not have enough money.

4 Check
Is my answer reasonable?

Problem Solving

Underline what you know. Circle what you need to find. Act it out to solve.

I'm going to be a pilot some day!

1. Jose has 1 quarter, 3 dimes, and 1 nickel. How much more does he need to buy an airplane that costs 75¢?

   _____ ¢

2. Leah has 1 quarter, 2 dimes, and a nickel. Her sister has 3 nickels. How many more cents do they need to have 90¢?

   _____ ¢

3. Cole has 75 cents. He has 2 quarters and 1 dime. If the rest of his coins are nickels, how many nickels does he have?

   _____ nickels

4. Sarah has 1 quarter, 1 dime, 3 nickels and 3 pennies. How much money does she have?

   _____ ¢

Math at Home Have your child show you the coins needed to buy a toy that costs 64¢.

Name

Dollars

Copyright © The McGraw-Hill Companies, Inc. (t)The McGraw-Hill Companies, (others)United States coin images from the United States Mint

Lesson 5

ESSENTIAL QUESTION
How do I count and use money?

Hi! I'm George.

Explore and Explain 🛠 Tools

one dollar = 100 cents

_____   _____   _____   _____

Teacher Directions: Count pennies to 100¢. Write the number of quarters it takes to equal 100¢. Do the same for the dime and each of the other coins.

See and Show

Dollar Sign → $1.00 ← Decimal Point

One **dollar** has a value of 100 cents or 100¢. To write one dollar, use a. **dollar sign.**

Use a decimal point to separate the dollars from the cents.

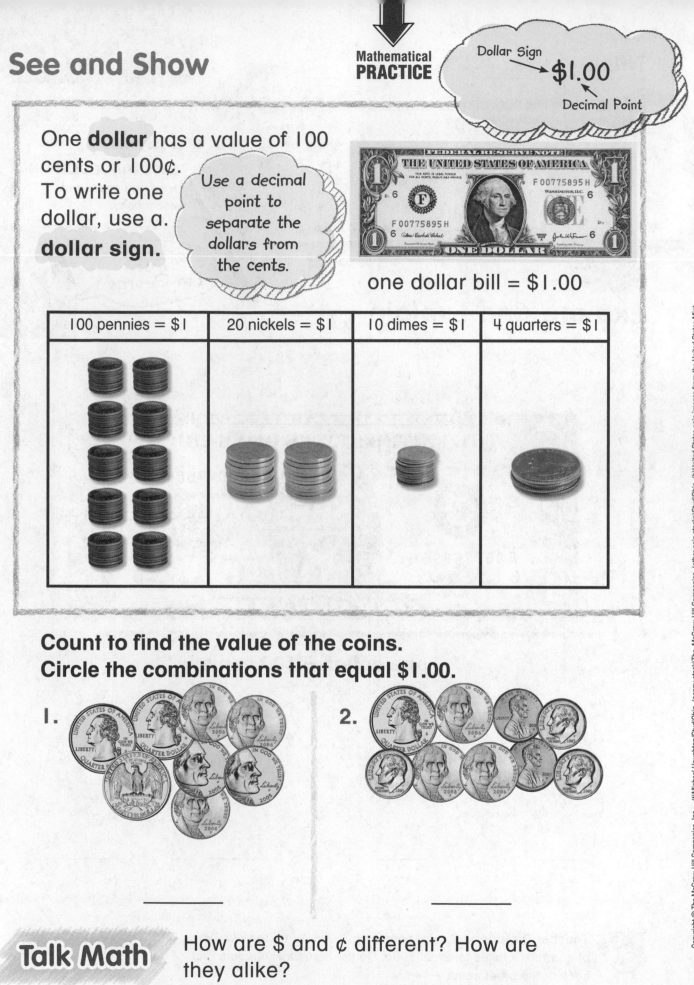

one dollar bill = $1.00

100 pennies = $1	20 nickels = $1	10 dimes = $1	4 quarters = $1

Count to find the value of the coins.
Circle the combinations that equal $1.00.

1.

2.

_____ _____

Talk Math How are $ and ¢ different? How are they alike?

Copyright © The McGraw-Hill Companies, Inc.   (t)Michael Houghton/StudiOhio, (penny stack)The McGraw-Hill Companies, (other coin stacks)StudiOhio, (b)United States coin images from the United States Mint

Name _____

On My Own

Count to find the value of the coins.
Circle the combinations that equal $1.00.

3.

_____

4.

_____

5.

_____

6.

_____

7.

_____

8.

_____

9. Natasha has 1 quarter, 2 dimes, 10 nickels, and 4 pennies. She needs 1 dollar to buy a joke book. How much money does she have?

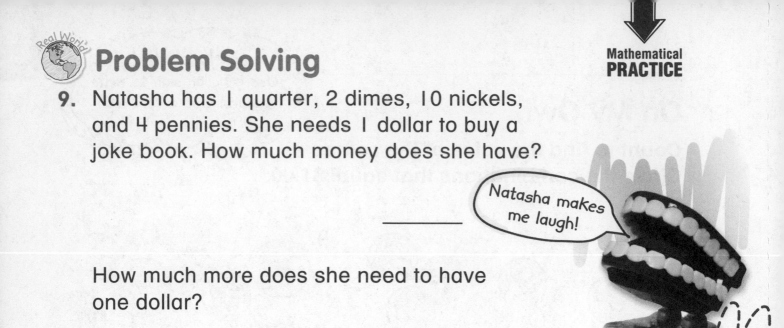

Natasha makes me laugh!

_____

How much more does she need to have one dollar?

_____

10. Chip needs 1 dollar. He has three quarters and one dime. How much does he have?

_____

How much more does he need to make 1 dollar?

_____

Write Math Think of 2 combinations of coins that equal one dollar and write them here.

_____

_____

_____

Measurement and Data
2.MD.8

CCSS

My Homework

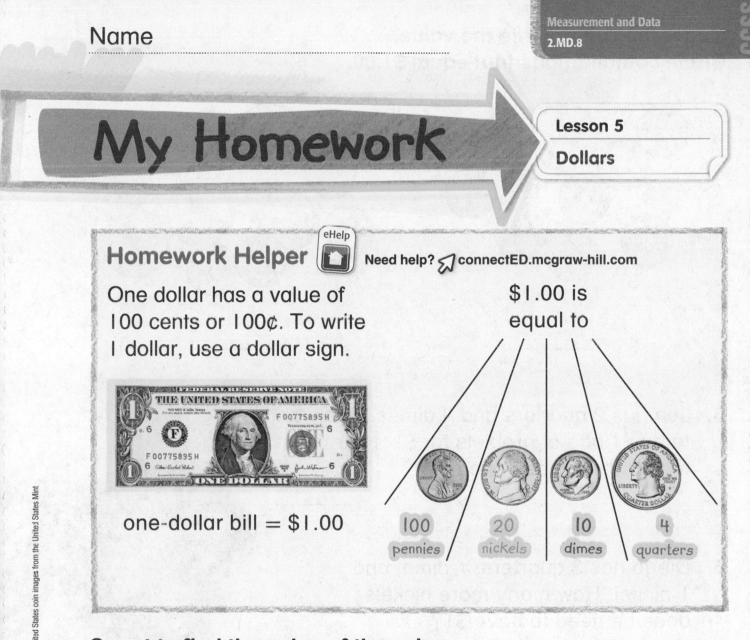

Homework Helper

Need help? connectED.mcgraw-hill.com

One dollar has a value of 100 cents or 100¢. To write 1 dollar, use a dollar sign.

one-dollar bill = $1.00

$1.00 is equal to

100 pennies / 20 nickels / 10 dimes / 4 quarters

Count to find the value of the coins.
Circle combinations that equal $1.00.

1. _____

2. _____

Count the coins. Write the value.
Circle combinations that equal $1.00.

3.

_____

4.

_____

5. Jen has 2 quarters and 4 dimes. She wants
   to buy 1 bag of pretzels for $1. How much
   more money does she need?

   _____

6. Diego has 3 quarters, 1 dime, and
   1 nickel. How many more nickels
   does he need to have $1?

   _____ nickels

Vocabulary Check [Vocab abc]

Circle the correct choices.

7. **one dollar** $1   1$ $1.00   1¢

Math at Home Have your child use various coins to show you two
ways to make $1.

Name

Vocabulary Check

Draw lines to match.

1. **dime**   1 cent or 1¢

2. **penny**   25 cents or 25¢

3. **quarter**   5 cents or 5¢

4. **dollar**   10 cents or 10¢

5. **nickel**   100 cents or 100¢

Concept Check

Count to find the value of the coins.

6.

_____¢, _____¢, _____¢, _____¢, _____¢ = _____¢

Concept Check

Count to find the value of the coins.

7.

_____¢, _____¢, _____¢, _____¢, _____¢, _____¢ = _____¢

Count to find the value of the group of coins.

8.

25¢   25¢   10¢   10¢   5¢   5¢

_____¢, _____¢, _____¢, _____¢, _____¢, _____¢

= _____¢

Count to find the value of the coins.
Circle the combinations that equal $1.00.

9.

_____

10.

_____

Use each coin to make one dollar.
Write the number of coins you used.

11.

_____

12.

_____

Problem Solving

I only cost 47¢.
I'm a bargain!

13. Lupe buys a toy dinosaur for 47¢.
He gives the cashier 1 quarter and
1 dime. How much more money
does he need to give the cashier?

_____

Circle the three coins Lupe should
give the cashier.

14. John has two dimes. Mark has two
quarters. How much money do the two
boys have in all?

_____

Test Practice

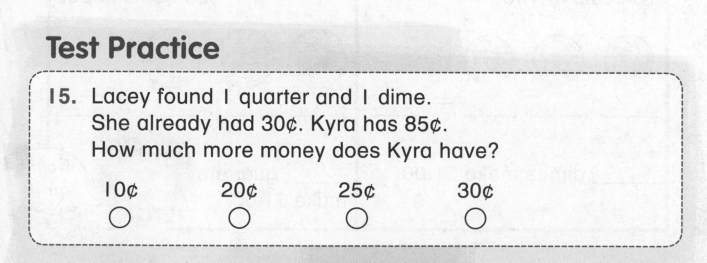

15. Lacey found 1 quarter and 1 dime.
She already had 30¢. Kyra has 85¢.
How much more money does Kyra have?

10¢	20¢	25¢	30¢
○	○	○	○

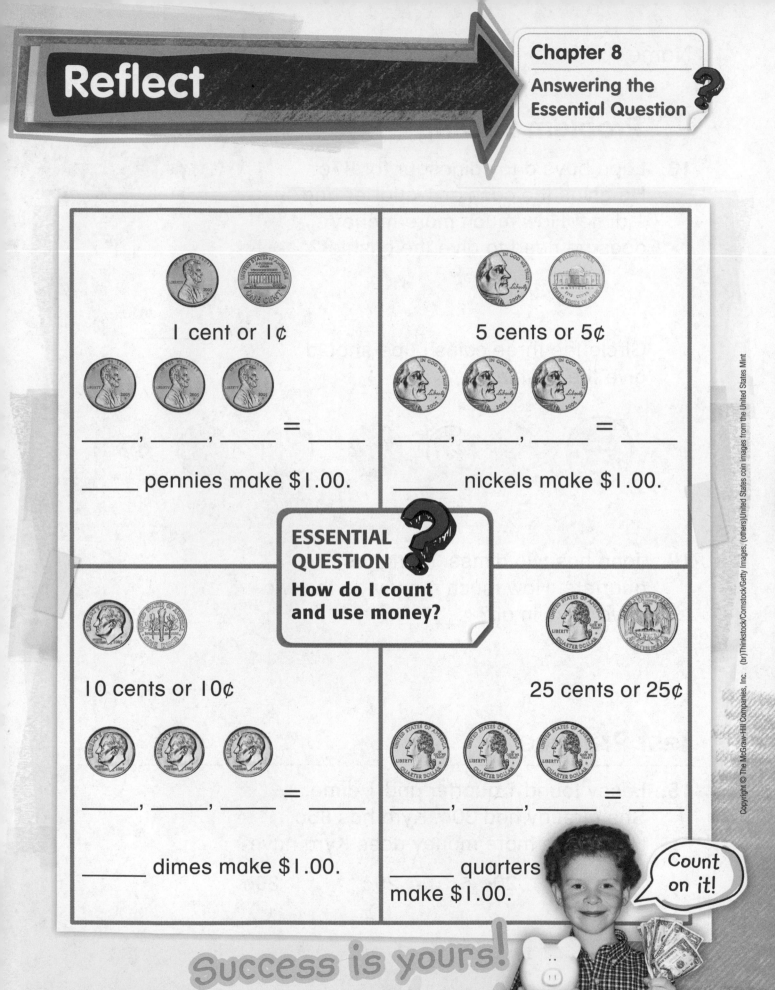

I cent or I¢

_____, _____, _____ = _____

_____ pennies make $1.00.

5 cents or 5¢

_____, _____, _____ = _____

_____ nickels make $1.00.

ESSENTIAL QUESTION
How do I count and use money?

10 cents or 10¢

_____, _____, _____ = _____

_____ dimes make $1.00.

25 cents or 25¢

_____, _____, _____ = _____

_____ quarters make $1.00.

Count on it!

Success is yours!

ESSENTIAL QUESTION

How can I record and analyze data?

Our Bodies Need Healthful Food!

Watch a video!

Watch

Measurement and Data

2.MD.9 Generate measurement data by measuring lengths of several objects to the nearest whole unit, or by making repeated measurements of the same object. Show the measurements by making a line plot, where the horizontal scale is marked off in whole-number units.

2.MD.10 Draw a picture graph and a bar graph (with single-unit scale) to represent a data set with up to four categories. Solve simple put-together, take-apart, and compare problems using information presented in a bar graph.

Standards for
Mathematical
PRACTICE ⬇

1. Make sense of problems and persevere in solving them.
2. Reason abstractly and quantitatively.
3. Construct viable arguments and critique the reasoning of others.
4. Model with mathematics.
5. Use appropriate tools strategically.
6. Attend to precision.
7. Look for and make use of structure.
8. Look for and express regularity in repeated reasoning.

= focused on in this chapter

Name **Eleanore**

Circle the group that has more.

Use the tally chart to answer the questions.

Favorite Pet	Tally
Cat	|||
Hamster	||||
Dog	⊬⊬ ||

2. What pet has 4 tally marks?

   Hampster

3. How many people like cats?

   3 people

Use the picture to solve.

4. Roy walks four dogs every day. How many brown dogs does he walk?

   1 brown dogs

How Did I Do? → Shade the boxes to show the problems you answered correctly.

My Math Words

Vocab
abc

Review Vocabulary

| compare | graph | tally marks |

Use the review words to describe each example.

What I Can Show on a Graph?

more →

fewer →

_____

How many balloons? ||||

_____

What is a graph?

_____

_____

_____

My Vocabulary Cards

Vocab abc

Mathematical PRACTICE

Lesson 9–4

bar graph

Swimming Laps

Lesson 9–1

data

What Pet Do You Like Best?

Pet	Tally
Dogs	
Cats	

Lesson 9–2

key

Favorite Vacation

Beach				
Camping				
Water park				

Key: Each picture = 1 vote

Lesson 9–7

line plot

Books Read in November

Lesson 9–2

picture graph

Our Favorite Toys

	Balls							
	Skates							
	Stuffed animals							

Lesson 9–1

survey

Favorite Time of Day		Tally	Total
Morning			6
Afternoon			3
Evening			8

Teacher Directions:
Ideas for Use
• Have students group 2 or 3 common
  words. Ask them to add a word that is
  unrelated to the group. Have them work
  with a friend to name the unrelated word.

• Have students write a tally mark on each
  card every time they read the word in this
  chapter or use it in their writing.

Numbers or symbols that
show information.

A graph that uses bars
to show data.

A graph used to show how
often a certain number occurs
in data.

Tells what, or how many,
each symbol stands for.

To collect data by asking
people the same question.

A graph that has different
pictures to show
information collected.

Lesson 9–2

symbol

Trees Planted	
Carla	🌲 🌲 🌲 🌲 🌲
Seth	🌲 🌲 🌲
Zane	🌲 🌲 🌲 🌲 🌲 🌲

Each 🌲 = 1 tree

symbol

Lesson 9–1

tally marks

Items Sold at School Store	
Item	Tally
Eraser	卌
Bottle of glue	卌 卌
Pencil	卌 III
Scissors	II

Teacher Directions:
More Ideas for Use
- Ask students to group like ideas they find throughout the chapter, such as using different ways to display data. Have students share the strategies they use to understand the concepts.

- Have students use a blank card to write this chapter's essential question. Have them use the back of the card to write or draw examples that help them answer the question.

Marks used in a survey to collect data.

A letter or figure that stands for something.

My Foldable

FOLDABLES® Follow the steps on the back to make your Foldable.

Name
....

Take a Survey

Lesson 1

ESSENTIAL QUESTION
How can I record
and analyze data?

Explore and Explain

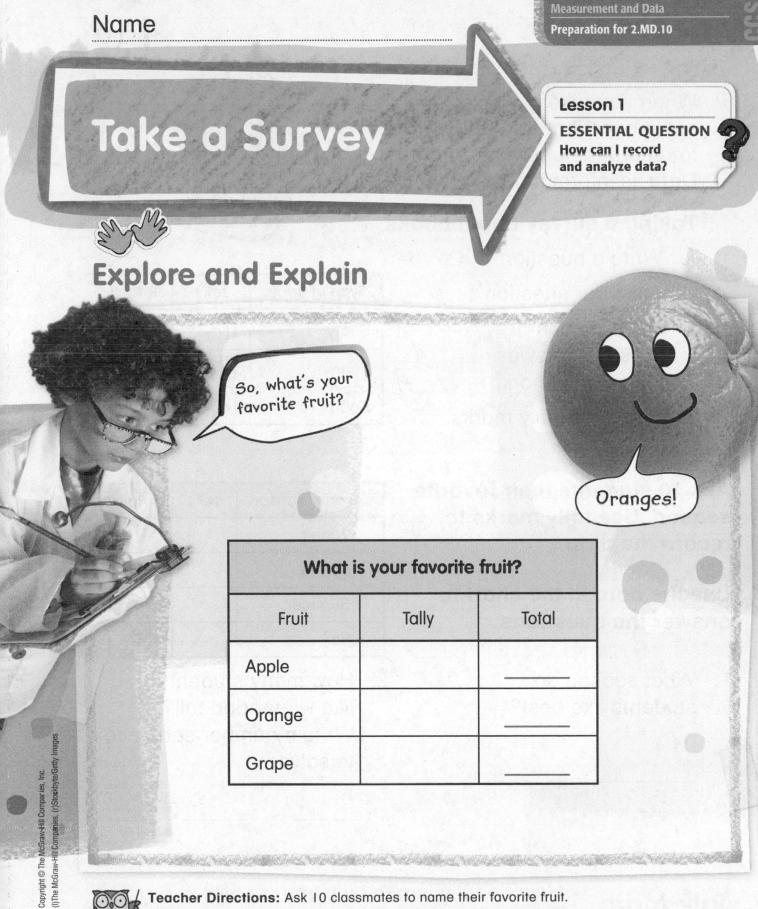

So, what's your favorite fruit?

Oranges!

What is your favorite fruit?		
Fruit	Tally	Total
Apple		_____
Orange		_____
Grape		_____

Teacher Directions: Ask 10 classmates to name their favorite fruit.
Mark a tally for each choice. Write the total for each fruit. Which fruit
do most people like?

Online Content at connectED.mcgraw-hill.com

Chapter 9 • Lesson 1

See and Show

When you take a **survey**, you ask a question. Use **tally marks** to record the answers, or data. **Data** is information.

I like scary books!

Taking a survey about books

1. Write a question.

2. Ask your question.

3. Record each person's answer with a tally mark.

4. Count the tally marks.

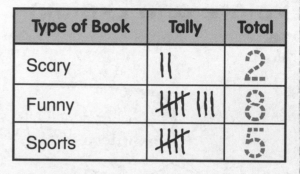

Type of Book	Tally	Total
Scary	II	2
Funny	IIII III	8
Sports	IIII	5

Ask 10 students their favorite season. Use tally marks to record the data.

Use the data in the chart to answer the questions.

Favorite Season	Tally	Total
Winter		
Spring		
Summer		
Fall		

1. What season do students like best?

   _____

2. How many students like winter and fall? Write a number sentence to solve.

   _____ ◯ _____ = _____

Talk Math How do tally marks help you organize data?

Name _____

On My Own

Ask 10 students their favorite picnic food. Use tally marks to record the data.

Helpful Hint

| means 1.

|||| means 5.

Picnic Food	Tally	Total
Hamburger		
Hot dog		
Chicken		

Use the data in the chart to answer the questions.

3. How many students like hamburgers? _____

4. How many students like hot dogs? _____

5. What is the favorite picnic food?

   _____

6. What is the least favorite picnic food?

   _____

7. How would the chart change if you added your

   favorite food? _____

   _____

Problem Solving

8. Claire took a survey of favorite snacks.
She marked 6 tallies for vegetables,
4 tallies for nuts, and 8 tallies for fruit.
How many friends did she survey?

Go nuts!

_____ friends

9. Students took a survey for Fun Friday
activities. Seven students want a picnic.
Five students want to go to the zoo.
Eight students want to watch a movie.
Which chart shows these results?
Circle the correct chart.

How many students were surveyed?

Activity	Tally
Picnic	卌 l
Zoo	卌
Movie	lll

Activity	Tally
Picnic	卌 ll
Zoo	卌
Movie	卌 lll

_____ students

HOT Problem How do you record data on
a tally chart?

_____

_____

_____

Name

My Homework

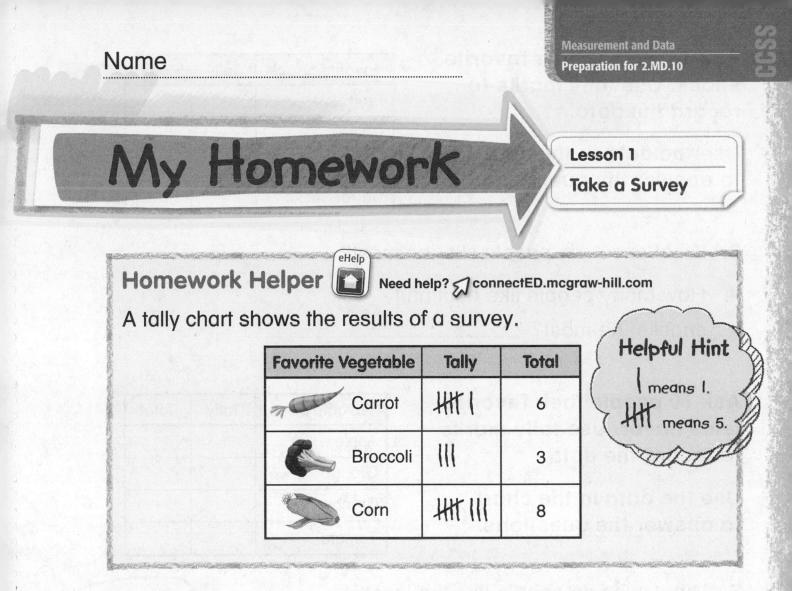

Homework Helper

eHelp

Need help? connectED.mcgraw-hill.com

A tally chart shows the results of a survey.

Favorite Vegetable		Tally	Total
Carrot		‖‖‖	6
Broccoli		‖‖‖	3
Corn		‖‖‖ ‖‖‖	8

Helpful Hint

| means 1.

‖‖‖ means 5.

Practice

Ask 10 people their favorite exercise. Use tally marks to record the data.

Use the data in the chart to answer the questions.

Favorite Exercise	Tally	Total
Jump Rope		
Run		
Dance		
Play Baseball		

1. How many people like to run and dance? _____

2. Do more people like to run and play baseball

   or dance and jump rope? _____

Copyright © The McGraw-Hill Companes, Inc.

Chapter 9 • Lesson 1   533

Ask 10 people their favorite snack. Use tally marks to record the data.

Use the data in the chart to answer the questions.

Favorite Snack	Tally	Total
Fruit		
Chips		
Cookies		
Carrot Sticks		

3. What snack do people like the most? _____

4. How many people like healthful snacks the most? _____

Ask 10 people their favorite juice flavor. Use tally marks to record the data.

Use the data in the chart to answer the questions.

Favorite Juice	Tally	Total
Apple		
Orange		
Cranberry		
Tomato		

5. What juice do people like the least? _____

6. How many people like cranberry juice and apple juice in all? _____

Wow! A lot of people like me!

Vocabulary Check

Vocab

Circle the word that matches the definition.

7. Numbers or symbols that show information.

   tally marks   survey   data

Math at Home Help your child to create a survey that they can give to family members.

Name

Make Picture Graphs

Explore and Explain

Lunchroom pals!

I love soup!

Favorite Lunch Food

Sandwich					
Pizza					
Hot Dog					
Soup					

Key: Each food = 1 vote

Teacher Directions: Ask five classmates to pick their favorite lunch food. Draw a picture in the chart to show each person's choice.

See and Show

You show data with a **picture graph**.
The pictures are a **symbol** for the data.

Favorite Marble Color

Blue	◯	◯	◯		
Red	◯	◯	◯		

Key: Each marble = 1 vote

Helpful Hint
The *key* tells how many
each *symbol* stands for.

Use the tally chart to make a picture graph.

1.

Favorite Shoes	Tally	Total
👟 Tennis Shoes	|||	3
🩴 Flip Flops	̶H̶H̶	5
👢 Boots	|	1

Favorite Shoes

Tennis Shoes					
Flip Flops					
Boots					

Key: Each shoe = 1 vote

Talk Math   How are picture graphs different
from tally charts?

Name

On My Own

Use each tally chart to make a picture graph.

2.

Favorite Drink	Tally	Total
Milk	|||	3
Apple Juice	||	2
Water	||||	4
Lemonade	||||	4

I'm not sticking out my tongue. That's just my straw!

Favorite Drink

Milk					
Apple Juice					
Water					
Lemonade					

Key: Each drink = 1 vote

3.

Favorite Sport	Tally	Total
Baseball	|||	3
Basketball	||||	4
Soccer	||	2
Football	||||	5

Favorite Sport

Baseball					
Basketball					
Soccer					
Football					

Key: Each ball = 1 vote

Problem Solving

Use the information to make a picture graph.

4. Lilian asked ten people their favorite flower. One said tulips. The same number of people said daisies and carnations. 5 said roses.

Favorite Flower					
Tulip					
Daisy					
Rose					
Carnation					

Key: Each flower = 1 vote

5. A school asked 15 students their favorite subject. 3 students said science and 3 said reading. One less person said art than math.

Favorite Subject					
Math					
Science					
Reading					
Art					

Key: Each book = 1 vote

Write Math   Explain how picture graphs can be more helpful than tally charts.

_____

_____

_____

Name

My Homework

Lesson 2

Make
Picture Graphs

Homework Helper eHelp

Need help? connectED.mcgraw-hill.com

You can use data from a tally chart to make a picture graph.

Favorite Milk	Tally	Total
Chocolate	⊮⊩	5
White	‖	2
Strawberry	‖‖	3

Favorite Milk

Chocolate	🥤	🥤	🥤	🥤	🥤
White	🥛	🥛			
Strawberry	🥤	🥤	🥤		

Key: Each glass = 1 vote

Practice

Use the tally chart to make a picture graph.

1.

Favorite Animal	Tally	Total
Elephant	‖‖	4
Giraffe	⊮⊩	5
Bear	‖‖	4
Snake	‖‖	3

Favorite Animal

Elephant					
Giraffe					
Bear					
Snake					

Key: Each picture = 1 vote

Mmm! Good and good for you!

Use the tally chart to make a picture graph.

2.

Favorite Cookie	Tally	Total
Chocolate Chip	||||	4
Peanut Butter	⊬|	5
Oatmeal Raisin	||||	4
Sugar	|||	3

Favorite Cookie					
Chocolate Chip					
Peanut Butter					
Oatmeal Raisin					
Sugar					

Key: Each cookie = 1 vote

Vocabulary Check

Circle the picture that matches the word.

3. picture graph

Type of Book	Tally	Total
Scary	||	
Funny	⊬| |||	
Sports	⊬|	

Favorite Pet					
Fish					
Dog					
Cat					

Key: Each animal picture = 1 vote

Number of Hats

```
        X
        X
      X X
    X X X
    X X X X
  X X X X X X
  +-+-+-+-+-+-+-
  0 1 2 3 4 5
```

  Math at Home Construct a large picture graph. Have your family graph their favorite dinner.

Name

Analyze Picture Graphs

Lesson 3
ESSENTIAL QUESTION
How can I record and
analyze data?

Explore and Explain

I Know
I am!

Favorite Dinner

Spaghetti						
Tacos						
Chicken						
Meatloaf						

Key: Each food = 1 vote

_____

_____

_____

_____

Which dinner is the least favorite? _____

Teacher Directions: Use the data in the picture graph. Write the number of votes
for each dinner. Write the least favorite dinner.

connectED.mcgraw-hill.com   Chapter 9 • Lesson 3   541

See and Show

You can use a picture graph to answer questions.

What pet is the favorite?

fish

Favorite Pet

Fish	🐟	🐟	🐟	🐟	🐟	🐟
Dog	🐶	🐶				
Cat	🐱	🐱	🐱	🐱		

Key: Each animal picture = 1 vote

How many votes does each picture show? _____ 1

Use the data from the graph to answer the questions.

Favorite Summer Activity

Swim	🏐	🏐	🏐	🏐	🏐
Bike Ride	🚲	🚲	🚲		
Roller Blade	👟	👟			
Play Baseball	⚾	⚾	⚾		

Key: Each picture = 1 vote

1. What activity is liked the least?

_____

2. Which two activities have the same number of votes?

_____

3. How many people like to swim and play baseball in all?

_____

Talk Math How would you count the votes for bike riding if each picture stood for 2 votes?

On My Own

Use the data from the graph to answer the questions.

How We Get to School						
Bus	🚌	🚌				
Walk	👟	👟	👟	👟	👟	
Bike	🚲					
Car	🚗	🚗				

Key: Each picture = 1 vote

4. How do most students get to school? _____

5. How many students ride the bus and ride in cars? _____

6. How many students voted? _____

7. How will the graph change if you add another way to get to school?

_____

8. Write a question that can be answered using the data in the picture graph.

_____

_____

9. Write a number sentence you can use to answer your question.

_____ ◯ _____ ◯ _____

Problem Solving

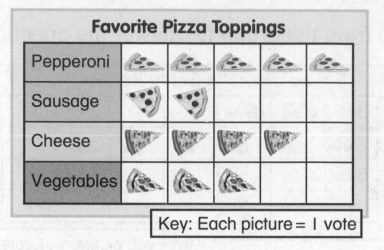

Use the data from the graph to answer the questions.

Favorite Pizza Toppings

Pepperoni	🍕	🍕	🍕	🍕	🍕
Sausage	🍕	🍕			
Cheese	🍕	🍕	🍕	🍕	
Vegetables	🍕	🍕	🍕		

Key: Each picture = 1 vote

10. Three more people were surveyed. They all like cheese. What topping is now the favorite?

_____

11. Carter voted for cheese. Kylie voted for vegetables. Whose favorite pizza topping got more votes?

_____

  Write a question about the graph above. Have a friend answer the question.

_____

_____

_____

Name

My Homework

Homework Helper

eHelp

Need help? connectED.mcgraw-hill.com

You can use the data from a picture graph to answer questions.

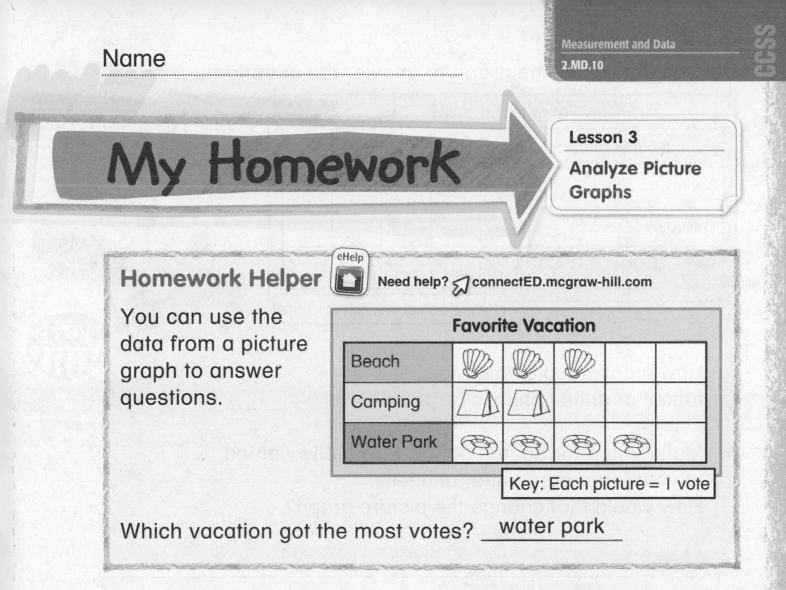

Favorite Vacation

Beach	🐚	🐚	🐚		
Camping	⛺	⛺			
Water Park	🛟	🛟	🛟	🛟	

Key: Each picture = 1 vote

Which vacation got the most votes? __water park__

Practice

Use the data from the graph to answer the questions.

Favorite Hamburger Topping

Ketchup	🍶	🍶	🍶	🍶	
Mustard	🍶	🍶			
Lettuce	🥬				
Tomato	🍅				

Key: Each picture = 1 vote

1. How many people voted?

   _____

2. Which topping got the most votes? _____

3. Which two toppings got the least votes? _____

4. How many people like ketchup or tomato? _____

Use the data from the graph to answer the questions.

Favorite Sandwich

Peanut butter and jelly	🥪	🥪	🥪		
Turkey	🥪	🥪			
Grilled cheese	🥪	🥪	🥪		
Peanut butter	🥪	🥪	🥪	🥪	

Key: Each sandwich = 1 vote

Vote for me! I'm for a longer lunch period!

VOTE

5. How many people voted for turkey or grilled cheese? _____

6. Maria voted for peanut butter. She really wanted to vote for peanut butter and jelly. How would that change the picture graph?

_____

_____

_____

Test Practice

7. In the graph above, how many more people like grilled cheese than turkey?

1	2	4	7
○	○	○	○

Math at Home Create a picture graph for your child about your family's favorite activities. Ask your child questions about the data on the graph.

Check My Progress

Vocabulary Check 🔤

Write the word to complete each sentence.

picture graph   **survey**   **key**

1. In a _____, you collect data by asking people the same question.

2. A _____ tells what (or how many) each symbol stands for.

3. A _____ has different pictures to show information collected.

Concept Check ✓

Ask 10 people their favorite color. Use tally marks to record the data.

Use the data in the chart to answer the questions.

Favorite Color	Tally	Total
Blue		
Pink		
Green		
Purple		

4. What color do people like the most? _____

5. How many people like pink and green? _____

6. What color do people like the least? _____

Use the tally chart to make a picture graph.

7.

Favorite Toy	Tally
Jump rope	|||
Baseball	||||
Roller blades	||
Bike	~~||||~~

Favorite Toy					
Jump rope					
Baseball					
Roller blades					
Bike					

Key: Each picture = 1 vote

Use the data from the graph to answer the questions.

8. How many more votes did bike get than baseball?

_____

9. How many people like jump ropes and roller blades?

_____

10. How many more people like to bike than jump rope?

_____

11. How many people like to jump rope and play baseball?

_____

Test Practice

12. Two students did not vote. If those 2 students all voted for roller blades, which toy would have the most votes?

Jump rope   Baseball   Roller blades   Bike
○   ○   ○   ○

Name _____

Make Bar Graphs

Lesson 4

ESSENTIAL QUESTION
How can I record and analyze data?

Explore and Explain ▶Watch

I ski.

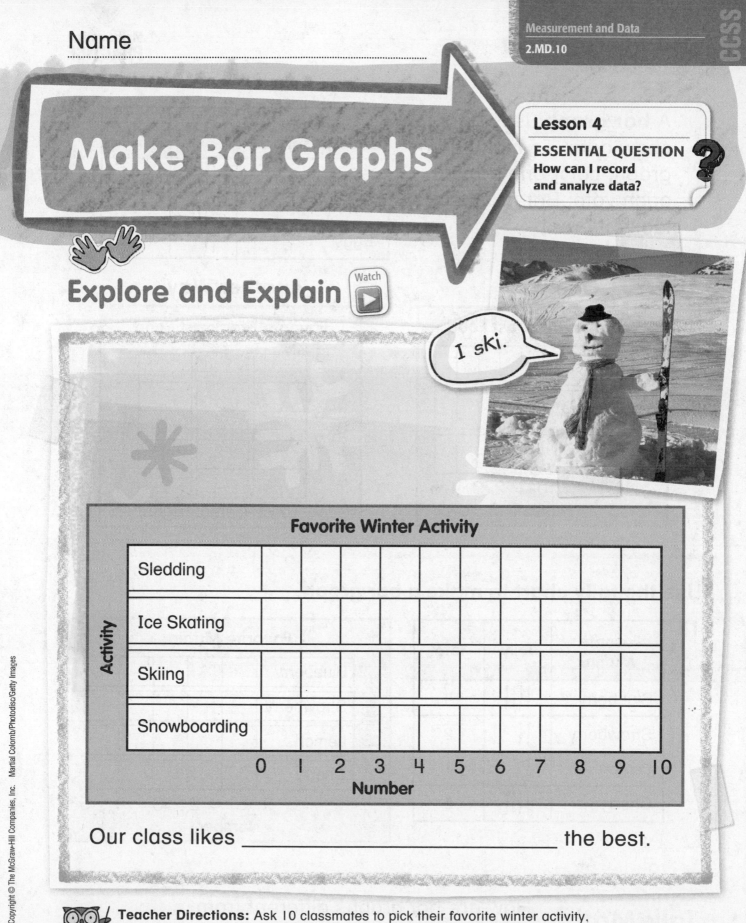

Favorite Winter Activity

Activity		0	1	2	3	4	5	6	7	8	9	10
Sledding												
Ice Skating												
Skiing												
Snowboarding												

Number

Our class likes _____ the best.

Teacher Directions: Ask 10 classmates to pick their favorite winter activity. Color one square for each vote in the correct row. Write your classmates' favorite activity.

See and Show

A **bar graph** uses bars to show data. To make a bar graph, color one box for each vote. Bar graphs can look different.

Favorite Breakfast Food	Tally	Total
Toast	||||	4
Cereal	|||||	5
Eggs	||	2

One Way

Favorite Breakfast Food

Food						
Toast						
Cereal						
Eggs						

0 1 2 3 4 5
Number

Another Way

Favorite Breakfast Food

Number
5 4 3 2 1 0

Toast Cereal Eggs

Use the tally chart to make a bar graph.

1.

Favorite Muffin	Tally	Total
Blueberry	||||	4
Strawberry	||	2
Lemon	|	1
Cherry	|||||	5

Favorite Muffin

Muffin						
Blueberry						
Strawberry						
Lemon						
Cherry						

0 1 2 3 4 5
Number

Talk Math How are bar graphs different from picture graphs?

Name _____

On My Own

Helpful Hint
Color 1 box
for each
tally mark.

**Complete the tally chart. Use the tally chart
to make a bar graph.**

2.

Color of Shirt	Tally	Total
Red	||||	
Blue	卌 |	
Black	||	
Green	卌	

Color of Shirt

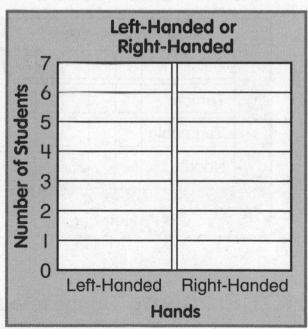

3.

Left-Handed or Right-Handed	Tally	Total
Left-Handed	||||	
Right-Handed	卌 |	

Problem Solving

Use the data to fill in the bar graphs.

4. Alyssa took a survey to find which juice pop flavor 15 friends like.
6 voted for cherry.
5 voted for grape.
2 voted for orange. The rest of her friends voted for lime.

Favorite Juice Pop

Flavor: Cherry, Grape, Orange, Lime

Number: 0 1 2 3 4 5 6 7

5. Luke took a survey of 18 friends to find their favorite soup. 4 voted for chicken soup.
6 voted for vegetable. The same number voted for tomato soup and noodle soup.

Favorite Soup

Soup: Chicken, Tomato, Vegetable, Noodle

Number: 0 1 2 3 4 5 6 7

It's as easy as ABC!

HOT Problem Suppose each square in the Favorite Soup graph above was worth 2 votes. How would your bar graph change?

_____

_____

Name _____

Measurement and Data
2.MD.10

CCSS

My Homework

Lesson 4

Make Bar Graphs

Homework Helper

eHelp

Need help? connectED.mcgraw-hill.com

You can use data from a tally chart to make a bar graph.

Favorite Music	Tally	Total
Country	|||	3
Rock	||||	4
Jazz	||	2

Favorite Music

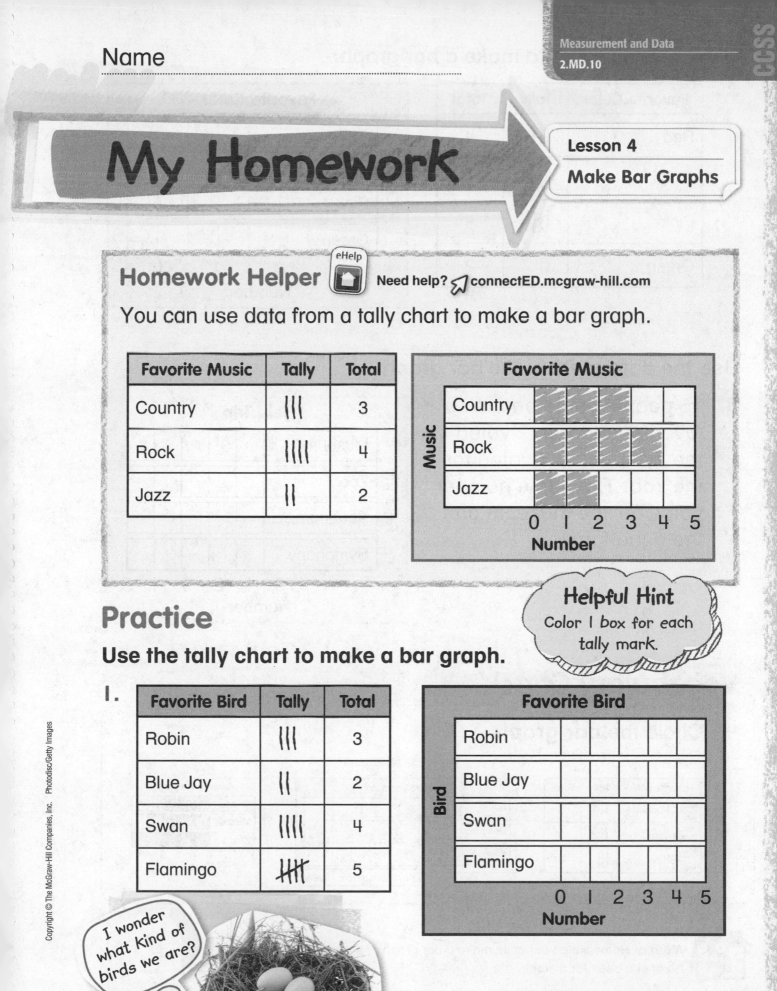

Practice

Use the tally chart to make a bar graph.

Helpful Hint
Color 1 box for each tally mark.

1.

Favorite Bird	Tally	Total
Robin	|||	3
Blue Jay	||	2
Swan	||||	4
Flamingo	~~||||~~	5

Favorite Bird

I wonder what kind of birds we are?

Use the tally chart to make a bar graph.

2.

Favorite Color	Tally	Total
Red	|||||	4
Blue	~~||||~~	5
Pink	|||||	4
Green	||	2

Favorite Color

Color

Red					
Blue					
Pink					
Green					

0 1 2 3 4 5

Number

Use the data to fill in the bar graph.

3. 15 people voted on their favorite field trip. 2 voted for the symphony. 5 voted for the zoo. The same number voted for the museum and the aquarium.

Field Trip

Field Trip

Museum					
Zoo					
Aquarium					
Symphony					

0 1 2 3 4 5

Number

Vocabulary Check 🔤

4. Circle the **bar graph**.

Favorite Milk	Tally	Total
Chocolate	~~||||~~	5
White	||	2
Strawberry	|||	3

Favorite Vacation

Beach	🐚 🐚 🐚	
Camping	⛺ ⛺	
Water park	🛟 🛟 🛟 🛟	

Key: Each picture = 1 vote

Favorite Topping

Topping

Onion					
Mustard					
Ketchup					

0 1 2 3 4 5

Number

 Math at Home Help your child make a bar graph of the type of weather he or she sees for a week.

Measurement and Data
2.MD.10

CCSS

Analyze Bar Graphs

Lesson 5

ESSENTIAL QUESTION
How can I record
and analyze data?

Explore and Explain

Healthful Dinner

Dinner	Number (0 to 10)
Chicken and Broccoli	
Pizza	
Spaghetti	
Milk and Cookies	

0 1 2 3 4 5 6 7 8 9 10

Number

chicken and broccoli _____

pizza _____

spaghetti _____

milk and cookies _____

Slurp!

Teacher Directions: Survey 10 people. Ask them which dinner is the healthiest. Make a bar graph. Write how many votes each dinner got.

See and Show

Owen surveyed his class about their favorite ice cream flavors. Then he made a bar graph to show the data.

Favorite Ice Cream Flavor	Tally
Chocolate	卌 II
Vanilla	III
Strawberry	卌
Mint	卌 I

Owen surveyed __21__ classmates.

Use the bar graph above to complete the sentences.

1. The flavor with the least votes is _____.

2. Chocolate received 2 more votes than _____.

3. What two flavors did eleven students vote for?

_____.

Talk Math   How could this graph help a teacher plan a class party?

Name

On My Own

Use the bar graph to answer the questions.

Hair Color

4. What color hair do 5 students have?

_____

5. What color hair do 4 students have?

_____

6. What two hair colors do the same number of students have?

_____

7. How many students were surveyed? _____

8. How would this graph change if 20 students took the survey?

_____

_____

9. Which pet do most students have?

_____

10. Which two pets do the fewest students have?

_____

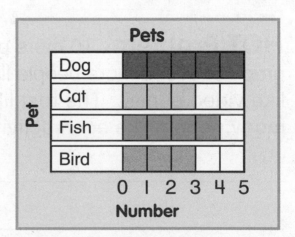

Pets

11. How many students have cats and birds? _____

Problem Solving

Use the bar graph to answer the questions.

Amazing! Samantha wins every time!

Favorite Activity

Activity						
Video games						
Playing outside						
Reading						
Board games						

0   1   2   3   4   5
Number

12. How many more people like reading or board games than playing video games?

_____

13. Samantha changes her vote from playing outside to reading. Which activity is the class favorite now?

_____

HOT Problem 10 more people were surveyed for the graph above. 3 more people like board games. 3 more people like video games. The rest like playing outside. Now how many people like each activity?

_____

_____

_____

Name _____

My Homework

Homework Helper

eHelp

Need help? connectED.mcgraw-hill.com

You can answer questions using the data from a bar graph.

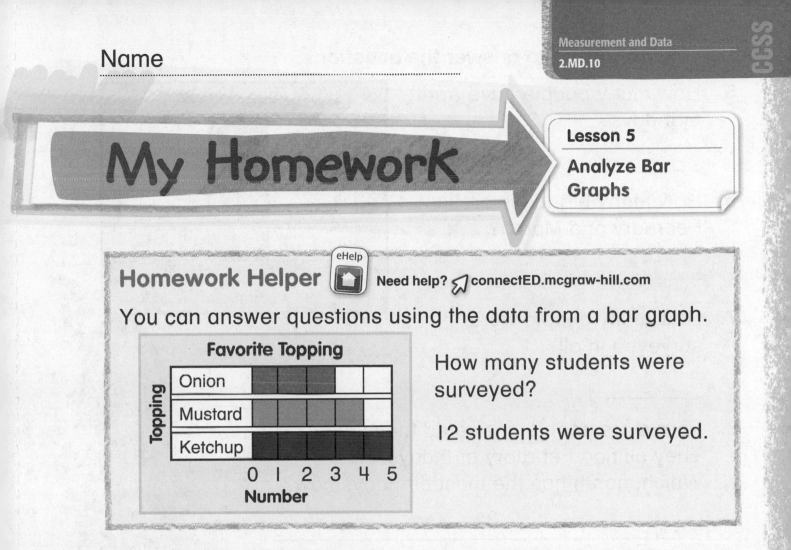

Favorite Topping

Topping: Onion, Mustard, Ketchup
Number: 0 1 2 3 4 5

How many students were surveyed?

12 students were surveyed.

Practice

Use the bar graph to answer the questions.

1. How many people voted for mustard?

   _____

2. What topping do most people like?

   _____

ketchup!

3. Which topping is liked the least?

   _____

4. How many people like mustard or onion? _____

Use the bar graph to answer the questions.

5. How many people have April birthdays?

   _____

6. How many birthdays are in February and March?

   _____

7. How many people were surveyed in all?

   _____

8. Four more people answered the survey. They all had February birthdays. Which month has the most birthdays now?

   _____

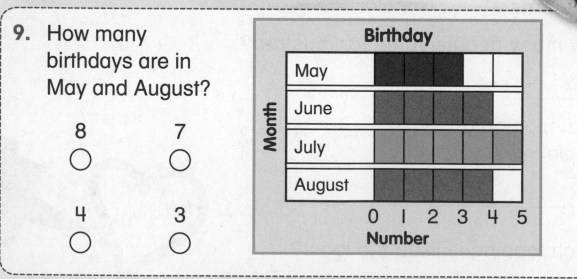

It's your big day!

Test Practice

9. How many birthdays are in May and August?

   8 ○   7 ○

   4 ○   3 ○

Math at Home Ask your child how he or she analyzed the bar graphs on this page.

Name _____

Problem Solving

STRATEGY: Make a Table

Lesson 6
ESSENTIAL QUESTION
How can I record and analyze data?

The breakfast special at Darla's Diner comes with 3 pancakes. Five friends order the breakfast special. How many pancakes will they get in all?

Watch ▶

Good morning!

1 Understand Underline what you know.
Circle what you need to find.

2 Plan How will I solve the problem?

3 Solve Make a table.

Friends	Pancakes
1	3
2	6
3	9
4	12
5	15

_15_ pancakes

4 Check Is my answer reasonable? Explain.

Practice the Strategy

Each person in the class has the same number of pets. Two people have 4 pets. How many pets do 7 people have?

1 Understand Underline what you know.
Circle what you need to find.

2 Plan How will I solve the problem?

3 Solve I will...

Number of Pets	
People	Pets
1	2
2	4
3	
4	
5	
6	
7	

_____ pets

4 Check Is my answer reasonable? Explain.

Squawk!

You've got this one!

Squawk!

Apply the Strategy

1. Desiree has 4 pairs of socks in her drawer.
   How many socks are there in all?

Pair	Socks in All
1	2

_____ socks

We're clean! Really!

2. Mr. Minnick needs to deliver 60 boxes.
   His car can hold 10 boxes at a time.
   How many trips will he need to make
   to deliver all 60 boxes?

_____ trips

3. Juice boxes come in packs of 4 at Sam's
   Grocery. Mrs. Perez needs 20 juice boxes
   in all. How many packs should she buy?

_____ packs

Review the Strategies

Choose a strategy
- Make a table.
- Make a model.
- Find a pattern.

4. Aaron has 3 muffin pans. Each pan can hold 6 muffins. How many muffins can Aaron bake at a time?

_____ muffins

5. Suzie, Simon, and Sean each had 4 bottles of water during their softball game. How many bottles of water did they have in all?

_____ bottles

6. Mr. Bell wants his 21 students to put on their gloves to go outside. How many gloves do they have in all?

_____ gloves

7. Grace, Elijah, and Cameron are saving soup labels. Grace has 10, Elijah has 7, and Cameron has 13. How many more labels does Cameron have than Elijah?

_____ labels

Name

My Homework

eHelp

Each hen lays the same number of eggs. I hen lays 3 eggs. Two hens lay 6 eggs. Three hens lay 9 eggs. How many eggs will 6 hens lay?

1 Understand Underline what you know. Circle what you need to find.

2 Plan How will I solve the problem?

3 Solve I will make a table.

Each hen lays 3 eggs. So, 6 hens will lay 18 eggs.

Hens	Eggs
1	3
2	6
3	9
4	12
5	15
6	18

4 Check Is my answer reasonable?

Problem Solving

**Underline what you know. Circle what you
need to find.**

1. Four turtles fit in one tank. Jose has 3 tanks.
   How many turtles can he have?

   _____ turtles

2. Nine children want to feed the birds. They each
   have 2 bags of seed. How many bags of seed
   are there in all?

   _____ bags

3. Shandra is giving a snack bag to each of her
   4 friends. She puts 4 pear slices in each bag.
   How many pear slices are there in all?

   _____ pear slices

   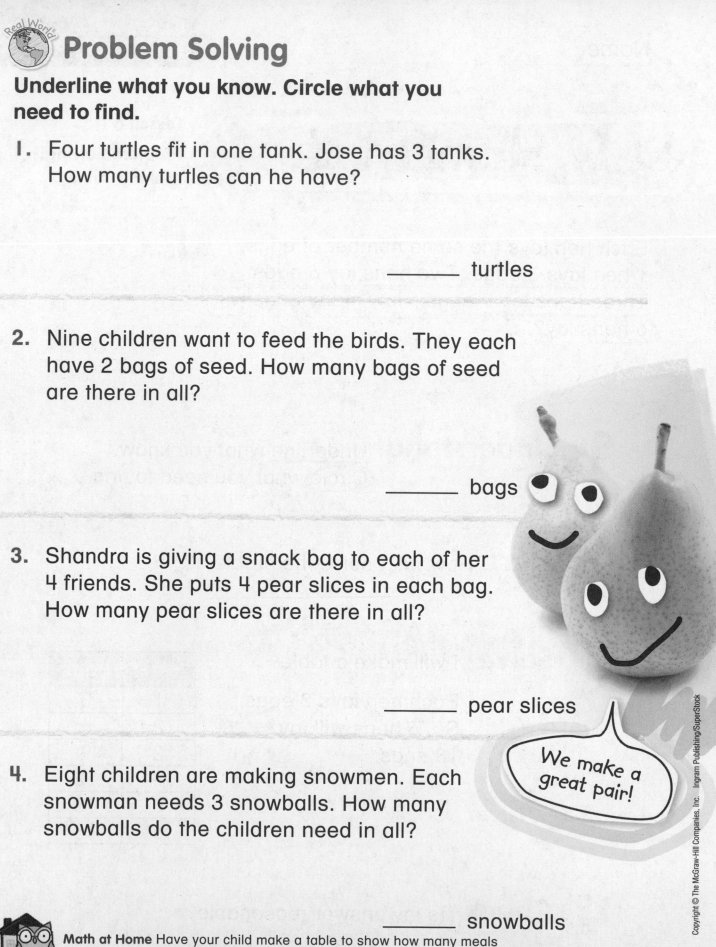

   We make a great pair!

4. Eight children are making snowmen. Each
   snowman needs 3 snowballs. How many
   snowballs do the children need in all?

   _____ snowballs

Math at Home Have your child make a table to show how many meals
he or she eats in a week.

Name

Make Line Plots

Lesson 7

ESSENTIAL QUESTION
How can I record and analyze data?

Explore and Explain

I'm ready to serve!

0 1 2 3 4 5 6 7 8 9 10

Teacher Directions: Chase asked 10 students how many servings of fruits and vegetables they eat each day. Three people eat 2 servings. Draw 3 X's over the number 2. Six people eat 5 servings. Draw 6 X's over the number 5. One person eats 7 servings. Draw 1 X over the number 7.

See and Show

A **line plot** is a way to organize data. Line plots are used to see how often a certain number occurs in data.

Mr. Sun's students marked their ages on a tally chart. They made a line plot using the data.

Age	Tally
7	|||
8	卌
9	||

Use the tally chart to make a line plot.

1.

Pets	Tally
1	||
2	|||
3	||
4	|

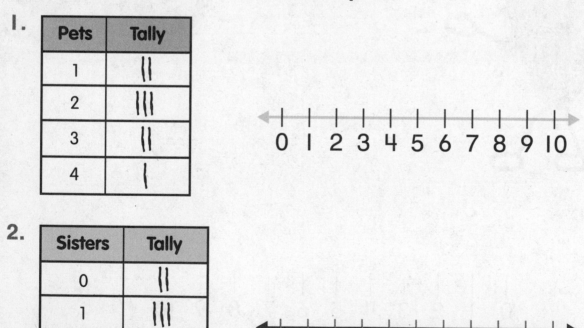

2.

Sisters	Tally
0	||
1	|||
3	|
4	

Talk Math   How are line plots similar to tally charts?

On My Own

Use the tally chart to make a line plot.

3.

Brothers	Tally
0				
1					
3				
4		

4.

Cousins	Tally
4				
5					
6	~~				~~	
7	~~||||~~

5.

Aunts	Tally
3	~~||||~~	
4				
5			
6				

We are like one BIG family in Mr. Sun's class!

Problem Solving

Use the data to make a line plot.

6. Eli asked 10 classmates how many glasses of water they drink each day. Two people said 1. Three people said 0. Three people said 4. Everyone else said 6.

0 1 2 3 4 5 6 7 8 9 10

7. Sydney asked 15 friends how many times they exercise each week. The same number of people said 1 time and 3 times. Five people said 2 times. Four people said 4 times.

0 1 2 3 4 5 6 7 8 9 10

Write Math   Can you use a line plot to show data about favorite color?

_____

_____

_____

My Homework

Homework Helper

eHelp   Need help? connectED.mcgraw-hill.com

You can use data from a tally chart to make a line plot.
Line plots show how often a certain number occurs in data.

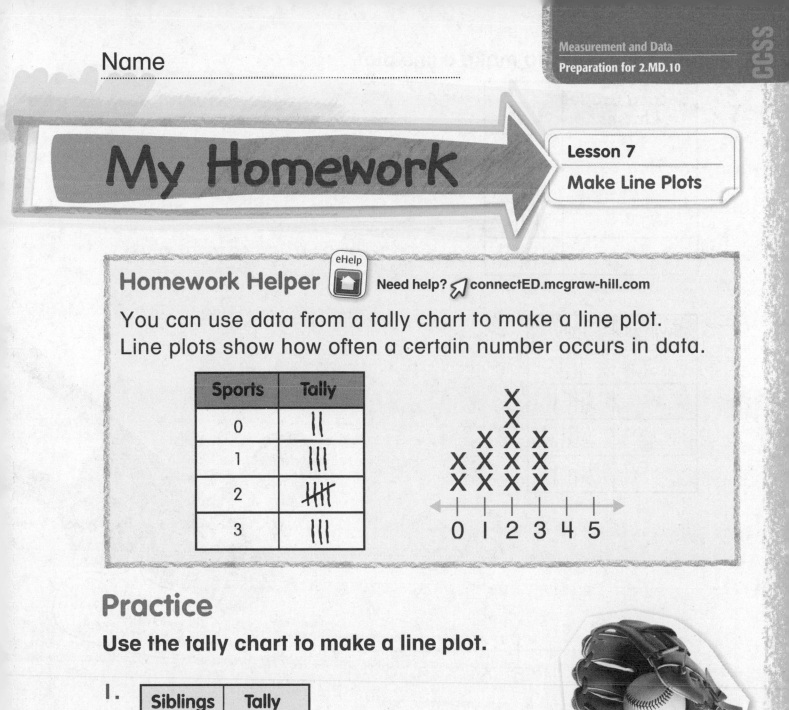

Sports	Tally
0	II
1	IIII
2	HHI
3	III

Practice

Use the tally chart to make a line plot.

1.

Siblings	Tally
0	IIII
1	I
2	III
3	II

Play ball!

Use the tally chart to make a line plot.

2.

Second Grade Classes	Tally
1	|
2	|||
3	̶H̶H̶
4	̶H̶H̶ ||

0 1 2 3 4 5 6 7 8 9 10

3.

Swings	Tally
5	|||||
6	̶H̶H̶ ||
7	|||
8	̶H̶H̶ |

0 1 2 3 4 5 6 7 8 9 10

I think I can touch this cloud with my toes!

Vocabulary Check

4. Circle the **line plot**.

Math at Home Help your child take a survey of the ages of his or her cousins. Have your child make a line plot to show the data.

Name _____

Analyze Line Plots

Lesson 8
ESSENTIAL QUESTION
How can I record and
analyze data?

Explore and Explain

Should we split dessert?

0 1 2 3 4 5 6 7 8 9 10

How many times a week do most people eat dessert? _____

Teacher Directions: Ask 10 people how many times they eat dessert each week. Use the data to make a line plot. Answer the question.

See and Show

You can use data from a line plot to answer questions.

This line plot shows the number of students in each class at a school.

Number of Students

How many students are in most of the classes? _____ 21

Use the data from the line plot to answer the questions.

1. How many students have 2 hats?

   _____

2. How many students have just 1 hat?

   _____

3. How many students have 3 hats?

   _____

4. How many students have more than 2 hats?

   _____

Number of Hats

Talk Math How are line plots the same or different than bar graphs or picture graphs?

Name

On My Own

Use the data from the line plot to answer the questions.

Number of Vegetables

Carrot buzz cut

5. Which number has the most Xs? _____

6. How many students like 4 vegetables? _____

7. Which number has the fewest Xs? _____

8. How many students like 2 vegetables? _____

9. How many students were surveyed in all? _____

10. How many students like more than 1 vegetable?

   _____ students

11. How many students do not like vegetables?

   _____ student

Problem Solving

12. Finish the line plot for Ashley.
She needs to show that three
friends each have 2 toy bears.

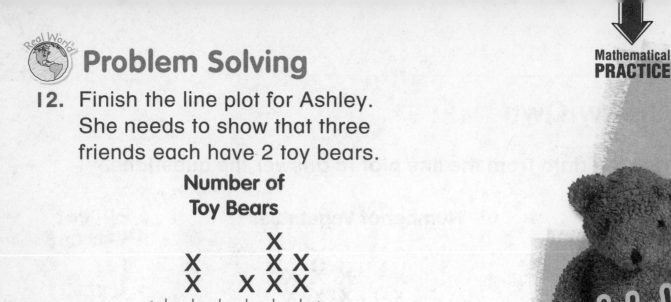

Number of
Toy Bears

13. 23 students created this line plot to show
how many fruits each student likes. How
many students do not like fruit at all?

Fruits

_____ students

Write Math   Colton looks at the line plot above and
says I student likes three fruits. Tell why
he is wrong. Make it right.

_____

_____

_____

My Homework

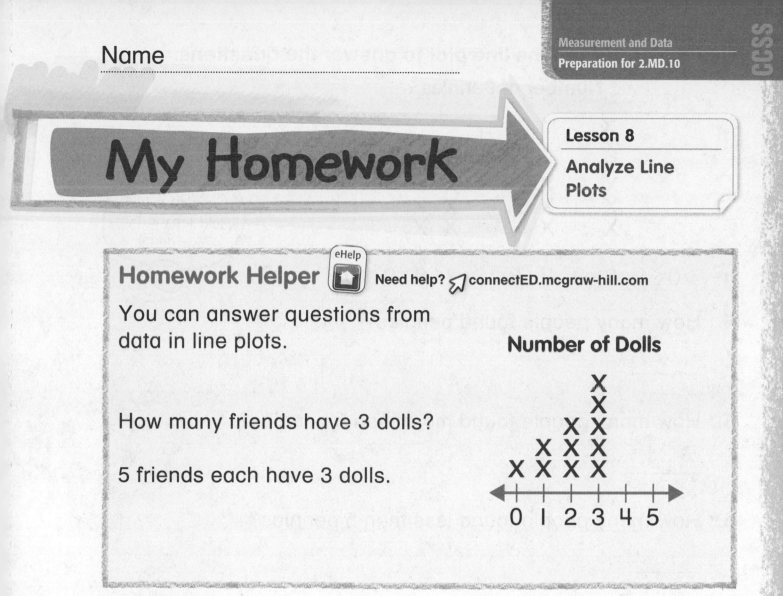

Homework Helper eHelp   Need help? connectED.mcgraw-hill.com

You can answer questions from data in line plots.

How many friends have 3 dolls?

5 friends each have 3 dolls.

Number of Dolls

Practice

Use the data from the line plot to answer the questions.

1. How many people get $5? _____

2. How much allowance do most people get?

   _____

3. How many people get more than $6?

   _____

My Friends' Allowances

$5 $6 $7 $8 $9 $10

Use the data from the line plot to answer the questions.

Number of Pennies

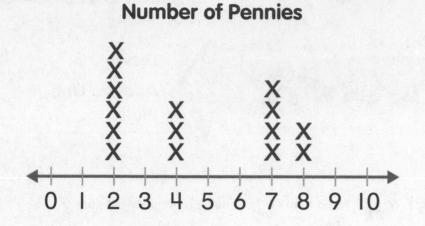

4. How many people found pennies?

   _____

5. How many people found more than 5 pennies?

   _____

6. How many people found less than 5 pennies?

   _____

Test Practice

7. How many people have 5 dogs?

   3   2
   ○   ○

   1   0
   ○   ○

Number of Dogs

Math at Home Make a line plot about the number of healthy food choices your child makes each day. Ask your child questions about the line plot at the end of a week.

578   Chapter 9 • Lesson 8

Name _____

My Review

Vocabulary Check

| data | survey | tally marks |
| picture graph | line plot | bar graph |

Write the correct word in each blank.

1. You can collect data by asking people the same question in a _____.

2. A _____ uses bars to show data.

3. Numbers or symbols that show information are called _____.

4. A _____ shows how often a certain number occurs in data.

5. A _____ uses pictures to show information collected.

6. A _____ is a mark used to record data collected in a survey.

Concept Check

Use the tally chart to make a picture graph.

7.

Favorite Sport	
Sport	Tally
Baseball	IIII
Football	IIII
Hockey	II
Volleyball	III

Favorite Sport

Baseball				
Football				
Hockey				
Volleyball				

Key: Each picture = I vote

Use the picture graph above to complete the sentences.

8. How many students were surveyed? _____

9. How many students like hockey and football? _____

Use the tally chart to make a bar graph.

10.

Have you been to the zoo?	Tally
Yes	HHT HHT
No	I

Have you been to the zoo?

Yes										
No										

0 I 2 3 4 5 6 7 8 9 10

Number of Students

Use the bar graph above to complete the sentences.

11. How many students have been to the zoo? _____

12. How many more students have been to the zoo than have not been to the zoo? _____

Name _____

Problem Solving

13. Taylor asked 12 friends to name their favorite number. Five friends said 4. Six friends said 1. One friend said 2. Use Taylor's data to complete the line plot.

Our Favorite Numbers

```
<----+--+--+--+--+--+--+--+--+--+--+---->
     0  1  2  3  4  5  6  7  8  9  10
```

My word problem isn't way up here. It's at the bottom of the page.

14. A tricycle has 3 wheels. How many wheels are there on 4 tricycles?

_____ wheels

Test Practice

15. There are 4 giraffes at the zoo. Each giraffe has 4 legs. How many legs are there in all?

     2   4   8   16

     ○   ○   ○   ○

Reflect

Show how to record and analyze data.

Complete the tally chart.

Fruit	Tally	Total
Banana	卌	
Apple	||||	
Orange	|||	

Use the tally chart to make a picture graph.

Fruit

Banana				
Apple				
Orange				

Key: Each picture = 1 vote

ESSENTIAL QUESTION

How can I record and analyze data?

Use the tally chart to make a bar graph.

Fruit

Fruit

Banana					
Apple					
Orange					

0 1 2 3 4 5
Number

Analyze the data.

How many people like apples and bananas?

_____

You can do it!

Keep your eyes on your goal!

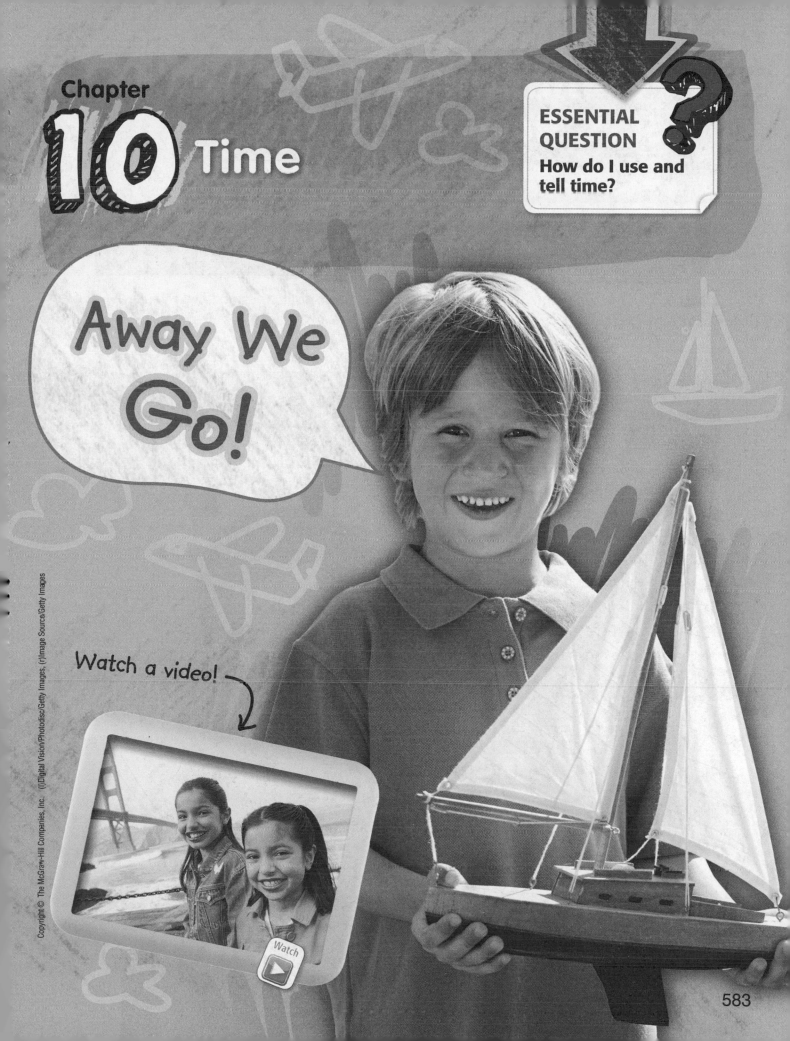

Chapter 10 Time

ESSENTIAL QUESTION

How do I use and tell time?

Away We Go!

Watch a video!

Watch

My Common Core State Standards

CCSS

Measurement and Data

2.MD.7 Tell and write time from analog and digital clocks to the nearest five minutes, using A.M. and P.M.

Standards for Mathematical PRACTICE ⬇

1. Make sense of problems and persevere in solving them.
2. Reason abstractly and quantitatively.
3. Construct viable arguments and critique the reasoning of others.
4. Model with mathematics.
5. Use appropriate tools strategically.
6. Attend to precision.
7. Look for and make use of structure.
8. Look for and express regularity in repeated reasoning.

💬 = focused on in this chapter

Name

Am I Ready?

Check ✓ ← Go online to take the Readiness Quiz

Circle the time.

1. 7:00   7 o'clock   2 o'clock   6 o'clock

2. 5:00   4:00   5:00   6:00

Find the missing numbers.

3.

0 1 2 3 4 5 ☐ 7 8 9 ☐ 11 12

4. 2, 4, _____, 8, _____, 12, 14, 16, _____, 20

5. 5, _____, _____, 20, _____, _____, 35, 40, _____

6. Kira is walking home from school. Circle the time of day for the activity.

morning   afternoon   evening

How Did I Do?

Shade the boxes to show the problems you answered correctly.

| 1 | 2 | 3 | 4 | 5 | 6 |

Name _____

My Math Words

Review Vocabulary

afternoon   evening   morning

Write about an activity you do for each time of day.
Use *morning*, *afternoon*, or *evening* in each sentence.

morning _____
_____
_____
_____

afternoon _____
_____
_____
_____

evening _____
_____
_____
_____

My Vocabulary Cards

Lesson 10–6

A.M.

Lesson 10–1

analog clock

Lesson 10–1

digital clock

5:00

Lesson 10–2

half hour

5:30

half past 5
5:30

Lesson 10–1

hour

3:00

hour

3 o'clock
3:00

Lesson 10–1

hour hand

A clock that has an hour hand and a minute hand.

The hours from midnight until noon.

One half hour is 30 minutes. Sometimes called half past.

A clock that uses only numbers to show time.

The shorter hand on a clock that tells the hour.

A unit to measure time. I hour = 60 minutes

Lesson 10–2

minute

1 minute
60 seconds

Lesson 10–1

minute hand

Lesson 10–6

P.M.

Lesson 10–4

quarter hour

12:15

12:45

quarter past 12   quarter till 1

Teacher Directions:
More Ideas for Use

• Tell students to draw a picture to show the meaning of a word. Have them work with a friend to guess the word.

• Have students use a blank card to write this chapter's essential question. On the back, have them write or draw examples that help them answer the question.

The longer hand on a clock that tells the number of minutes.

A unit to measure time. Each tick mark on an analog clock.
I minute = 60 seconds
60 minutes = I hour

A quarter hour is 15 minutes. Sometimes called quarter past or quarter till.
4 quarter hours = I hour

The hours from noon until midnight.

My Foldable

A.M.   **P.M.**

A.M.   **P.M.**

A.M.   **P.M.**

A.M.   **P.M.**

FOLDABLES® Study Organizer

1 o'clock

quarter till 9

quarter past 2

half past 3

Name

Time to the Hour

Lesson 1
ESSENTIAL QUESTION
How do I use and
tell time?

Explore and Explain

This is a cool way to get to school!

I hope we're not late.

_____ o'clock

Teacher Directions: Write the missing numbers on the clock.

Use 🕐 to show time to each hour. Draw hands on the clock to show 4 o'clock. Write the time.

See and Show

On an **analog clock**, the **hour hand** points to the **hour**. It is shorter.

A **digital clock** shows the hour and minutes on a screen.

The **minute hand** points to the **minute**. It is longer.

minute hand

hour hand

3:00

↑ hour   ↑ minutes after the hour

_____ o'clock

Use . Tell what time is shown. Write the time.

1.

_____ o'clock

2.

_____ o'clock

3.

_____ o'clock

Talk Math   How is an analog clock similar to a number line?

Name

On My Own

Use . Tell what time is shown. Write the time.

4.

_____ o'clock

5.

_____ o'clock

6.

_____ o'clock

Draw the hands on the clock. Write the time.

7. 5 o'clock

8. 10 o'clock

9. 3 o'clock

Problem Solving

Show the time on each clock.

10. Colin gets home at 3 o'clock. Evan gets home one hour later. What time does Evan get home?

11. Basketball practice starts at 4:00. Mark eats a snack 2 hours before practice. What time does Mark eat his snack?

12. It is after 7 o'clock and before 9 o'clock. What time is it if it is exactly on the hour? Show the time on the analog clock face.

Write Math What time is it when the hour hand is on 5 and the minute hand is on 12? How do you know?

_____

_____

_____

My Homework

Homework Helper eHelp 🏠 Need help? connectED.mcgraw-hill.com

You can tell and write time to the hour.

minute hand

hour hand

It is 5 o'clock or 5:00.

hour minutes after the hour

Practice

Tell what time is shown. Write the time.

1.

_____ o'clock

2.

_____ o'clock

3.

_____ o'clock

Draw the hands on each clock. Then write the time.

4. 2 o'clock

5. 4:00

6. The time on Javier's watch is 3 hours after 2:00. Write the time on his watch.

Javier's cool watch!

Vocabulary Check

Complete each sentence.

hour hand   analog clock   minute hand   digital clock

7. An _____ has an hour and a minute hand.

8. The _____ points to the hour on an analog clock.

9. A _____ uses numbers on a screen to show time.

10. The _____ points to the minute on an analog clock.

Math at Home Throughout the day, ask your child to look at an analog or digital clock showing a time to the hour and have them tell you the time.

Name

Time to the Half Hour

Lesson 2

ESSENTIAL QUESTION
How do I use and tell time?

Explore and Explain

Watch ▶

Tools

It's time for our trip!

half past _____

Teacher Directions: Use a 🕐 to show 5 o'clock. Move the minute hand to 6.
What time is it? Repeat with other times showing hour and half past the hour.
Draw hands to show one of the times and write the time.

You can show time to the half hour.
One half hour is 30 minutes.
These clocks show **half past** 2.

Helpful Hint
The hour hand moves halfway to the next number as a half hour passes.

It is 2:30.

The hour hand points between 2 and 3. The minute hand points to 6.

Use 🕐. **Tell what time is shown. Write the time.**

1.

half past _____

2.

half past _____

3.

half past _____

Talk Math   It is half past 8. Explain what *half past* means.

Name _____

On My Own

Use . Tell what time is shown. Write the time.

4.

half past _____

5.

half past _____

6.

half past _____

Draw the hands on each clock. Write the time.

7. half past 4

8. half past 7

9. half past 6

Problem Solving

10. Eli woke up at half past 7. He needs to leave for school one hour later. Show and write the time he needs to leave for school.

Use the clocks shown below.

School starts   Soccer practice   Alarm goes off

What happens at each of these times?

11. 6:30 _____

_____

12. 3:30 _____

_____

13. 8:30 _____

Write Math   How is reading a digital clock different from reading an analog clock? Explain.

_____

_____

_____

My Homework

Homework Helper eHelp Need help? connectED.mcgraw-hill.com

You can tell and write time to the half hour.

hour hand

minute hand

hour minutes after hour

It is half past 3 or 3:30.

Practice

Tell what time is shown. Write the time.

1.

half past _____

2.

half past _____

3.

half past _____

Draw the hands on each clock. Write the time.

4. half past 1

5. half past 11

6. Tatum is going to the city with her family. They will leave one hour after she gets home from school. Tatum gets home from school at 3:30. What will the clocks look like when they leave?

No, we're not lost. We're on an adventure!

Vocabulary Check

Show a **half hour** later on the second clock.

7.

 Math at Home Give your child a time to the hour. Have him or her tell you the position of the clock hands for half past that hour or in a half hour.

Problem Solving

STRATEGY: Find a Pattern

Lesson 3

ESSENTIAL QUESTION
How do I use and tell time?

The buses leave in order every half hour. Bus 1 leaves at 9:30. Bus 2 leaves at 10:00. What time do buses 3 and 4 leave?

Tools

All aboard!

1 Understand   Underline what you know.
Circle what you need to find.

2 Plan   How will I solve the problem?

3 Solve   Find a pattern.

$\underline{\quad 9:30 \quad}$, $\underline{\quad 10:00 \quad}$, $\underline{\quad 10:30 \quad}$, $\underline{\quad 11:00 \quad}$

Bus 1   Bus 2   Bus 3   Bus 4

4 Check   Is my answer reasonable?

Practice the Strategy

Each class at school will go to recess 30 minutes after the other. The first class goes to recess at 1:00. What time will the fourth class go?

I'm going to the moon!

Blast off!

1 Understand Underline what you know.
Circle what you need to find.

2 Plan How will I solve the problem?

3 Solve I will...

————, ————, ————, ————

Class 1   Class 2   Class 3   Class 4

4 Check Is my answer reasonable? Explain.

Name _____

Apply the Strategy

Find a pattern to solve. Use the space provided.

1. In the morning, Ms. White's students change activities every hour. Reading starts at 8:30. Her students go to three more activities. When do each of the activities start?

   8:30, _____, _____, _____

2. Students start work in learning stations at 10:30. After two hours, they go to recess. What time do students go to recess?

   _____

3. In the afternoon, Ms. White's students change activities every half hour. Math starts at 1:00. Writing starts after math. What time will writing start?

   _____

4. Carmen has 8 shoes. She places an equal number of shoes in two suitcases. How many shoes will be in each suitcase?

_____ shoes

5. A postcard costs 45¢. Jason wants to buy 2 postcards. He has 2 quarters, 2 dimes, and 3 nickels. How much money does he have?

_____

How many pennies does he need to buy the postcards?

_____ pennies

6. The subway arrives at the station on the hour and half hour. It is now 4:30. Kate's family is on the way but it will take one hour to get there. What is the earliest time her family can get on the subway?

_____

Name _____

My Homework

Lesson 3
Problem-Solving:
Find a Pattern

Ashley's family is going to visit their grandma. It will take them four hours to get there. They leave at 10:30. What time will they get to grandma's house?

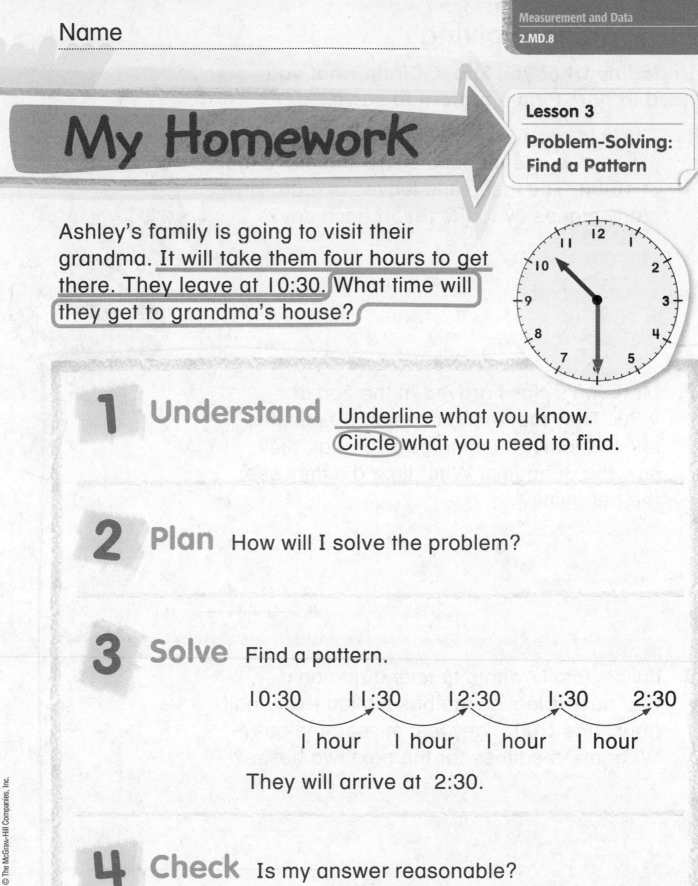

1 Understand Underline what you know.
Circle what you need to find.

2 Plan How will I solve the problem?

3 Solve Find a pattern.

10:30   11:30   12:30   1:30   2:30

I hour   I hour   I hour   I hour

They will arrive at 2:30.

4 Check Is my answer reasonable?

Problem Solving

**Underline what you know. Circle what you
need to find. Find a pattern to solve.**

1. There is one plane that leaves from the city to
   the ski resort every hour. The first plane leaves
   at 10:00. The last plane leaves at 4:00. How
   many planes fly to the resort each day?

   _____ planes

2. Mr. Lyon's class arrived at the zoo at
   9:30. They met the zookeeper 30 minutes
   later. Another 30 minutes after that, they
   saw the penguins. What time did they see
   the penguins?

   _____

3. Julio's family wants to take a ride on a
   tour bus. It leaves the bus station every half
   hour. It is 1:00. They just missed the bus.
   What are the times for the next two buses?

   _____ , _____

Math at Home Ask your child to keep a journal for one evening. At each half hour,
have your child write the time and record his or her activity at that time. At the end
of the evening, see if there were any repeating or other patterns in their activities.

Check My Progress

Vocabulary Check

Complete each sentence.

minute   analog clock   hour

digital clock   half hour

1. 60 minutes is one _____.

2. 30 minutes is one _____.

3. 60 seconds is one _____.

4. A _____ uses only numbers to show time.

5. An _____ has an hour hand and minute hand to show time.

Concept Check

Tell what time is shown. Write the time.

6.

7.

8.

Draw the hands on each clock. Write the time.

9. half past 8

10. 11 o'clock

11. half past 1

Write the time.

12.

7:30

half past _____

13.

1:00

_____ o'clock

14. Kevin has three classes in a row. Each class is 2 hours long. Kevin's first class is at 7:00. When will Kevin's last class end?

_____

Test Practice

15. The class bell rings every 30 minutes. The first one rings at 8:30. What time will the 4th bell ring?

10:30   10:00   11:30   9:30
○   ○   ○   ○

Name _____

Time to the Quarter Hour

Lesson 4

ESSENTIAL QUESTION
How do I use and tell time?

Explore and Explain

Watch ▶ Tools

I can lend a hand with quarter hours!

quarter past _____

Teacher Directions: Have students fold a paper plate, or circular piece of paper in half. Fold it again in quarters. Unfold and look at the fold marks. On the clock above, place cubes on the quarter hours referring back to their plate or circle. Draw hands on the clock to show quarter past an hour. Write the time.

See and Show

You can show time to the **quarter hour**. There are 15 minutes in a quarter hour. There are 4 quarter hours in an hour.

| 1 o'clock | quarter past 1 | half past 1 | quarter till 2 |

Use 🕐 **. Tell what time is shown. Write the time.**

1.

2.

3.

Talk Math   At 4:15, where is the minute hand? Explain.

Name _____

On My Own

Use . Tell what time is shown. Write the time.

4.

5.

6.

Draw the hands on each clock. Write the time.

7. quarter till 2

8. quarter past 5

9. quarter past 12

Read the time. Write the time on the digital clock.

10. twelve forty-five

11. four fifteen

12. nine thirty

Problem Solving

13. Mia is at school. School is over at 3:15. She has 3 more hours to wait. What time is it now?

_____

14. Ken goes to his friend's house at 2:15. Circle the clock that shows this time.

15. Alea's family went hiking. They left at 9:45. They drove for 3 hours. Then they stopped for lunch. Lunch took one hour. What time did they finish lunch?

_____

Follow me!

HOT Problem Why is each 15-minute period on a clock called one quarter hour?

_____

_____

_____

My Homework

Homework Helper eHelp Need help? connectED.mcgraw-hill.com

You can show time to the quarter hour.

quarter past 8 | half past 8 | quarter till 9

8:15 | **8:30** | **8:45**

Practice

Use 🕐. Tell what time is shown. Write the time.

1.

2.

3.

Draw the hands on each clock. Write the time.

4. quarter past 6   5. quarter till 3   6. quarter past 7

7. Carly left for school at 8:15. She left school at 2:45. If she arrived at school at 8:45, how many hours was she in school?

The day flew by!

_____ hours

Vocabulary Check

8. Circle the clock that shows an example of quarter hour.

12:15   7:35

Math at Home Have your child use the words quarter till and quarter past to describe the time at 6:15 and 6:45.

Time to Five Minute Intervals

Lesson 5

ESSENTIAL QUESTION
How do I use and tell time?

Explore and Explain

Teacher Directions: Count by 5s as you trace the dotted line. Label each jump 5, 10, 15 and so on. Draw the minute hand. Write that time on the digital clock.

Mathematical PRACTICE

It takes 5 minutes for the minute hand to move to the next number. You can skip count by 5s to tell the time.

This clock shows ___40___ minutes after 9 o'clock.
Write the time another way.

Helpful Hint
Each mark on the clock face is one minute.

`9:40`

**Tell what time is shown. Use 🕐 to help.
Write the time.**

1.

2.

3.

4. Tell what time is shown. Draw the minute hand to show the time.

`10:35`

Talk Math Explain how you skip count by 5s to tell time.

Name

On My Own

Tell what time is shown. Use to help. Write the time.

5.

6.

7.

Tell what time is shown. Draw the hands to show the time.

8.
5:45

9.
12:35

10.
9:50

11. 2:20

12. 8:45

13. 1:25

14. If the hour hand is close to the 11 and the minute hand is pointing to the 10. What time is it?

_____

15. A group of people get on an amusement park ride every 5 minutes. It is now 3:00. There are 7 groups ahead of Alyssa and her family. What time will her family go on the ride?

I can see my house from here!

16. It takes 20 minutes to watch the movie previews. The previews start at 7 o'clock. What time will the movie start?

Write Math What time is it when the hour hand is between 5 and 6 and the minute hand is on 7? Explain.

_____

_____

_____

Name _____

My Homework

Lesson 5

Time to Five-Minute Intervals

Homework Helper

Need help? connectED.mcgraw-hill.com

It takes 5 minutes for the minute hand to move to the next number. Skip count by 5s to tell time.

This clock shows 35 minutes after 4 o'clock.

The time is shown another way.

Practice

Read the time. Write the time.

1.

2.

3.

Tell what time is shown. Draw the minute hand to show the time.

4.

`11:35`

5. `7:05`

6. `3:25`

7. It is 1:00. Hunter is waiting for Ben. Ben said he would meet Hunter in 25 minutes but he arrived 5 minutes late. What time does Ben meet Hunter?

Test Practice

8. A train leaves at 10:40. Which clock shows 10:40?

○   ○   ○   ○

Math at Home Ask your child to look at an analog clock on the hour. Have him or her tell you what time it will be in 5 minutes, 10 minutes, 25 minutes, and 50 minutes.

Name

A.M. and P.M.

Lesson 6

ESSENTIAL QUESTION
How do I use and tell time?

Explore and Explain

Whoo hoo!

Where do you dream of traveling to?

Teacher Directions: Think of things that you do during the day and at night. Draw pictures of the activities on the correct side of the page.

See and Show

The hours from midnight until noon are labeled **A.M.** The hours from noon until midnight are labeled **P.M.**

> **Helpful Hint**
> 12:00 P.M. is noon.
> 12:00 A.M. is midnight.

Go to School

8:00 A.M.

Read a Bedtime Story

8:00 P.M.

Tell what time is shown for the activity. Write the time. Circle A.M. or P.M.

I. Art class

   A.M.
   P.M.

2. Go to Bed

   A.M.
   P.M.

3. Play after School

   A.M.
   P.M.

Talk Math How can you remember if it is A.M. or P.M.?

On My Own

Tell what time is shown for the activity. Write the time.
Circle A.M. or P.M.

4. Eat Breakfast

A.M.
P.M.

5. Wash the Dog

A.M.
P.M.

6. Go Swimming

A.M.
P.M.

Tell what time is shown for the activity. Draw the
hands on the clock to show the time. Circle A.M. or P.M.

7. Soccer Practice

3:30

A.M.
P.M.

8. Sleeping in Bed

2:30

A.M.
P.M.

9. Eat Dinner

6:45

A.M.
P.M.

Complete each sentence. Write *midnight* or *noon*.

midnight   noon

10. At _____, I am usually asleep.

11. At _____, I may be eating my lunch.

12. Zack has lacrosse practice at 4:00. Would that likely be at 4:00 A.M. or 4:00 P.M.?

_____

Library time is my favorite time of day!

13. It is 2:25. Lakota is going to the library in 1 hour. On the clock show the time she will go to the library.
Circle A.M. or P.M.

A.M.   P.M.

14. It is 3:00 A.M. Are you more likely to be on your way home from school or sleeping in your bed?

_____

 Write Math   Christy wants to go to her friend Jill's party. The invitation says the party starts at 1:30 A.M. This is wrong. What mistake was made on the invitation?

_____

_____

My Homework

Homework Helper

Need help? connectED.mcgraw-hill.com

The hours from midnight until noon are labeled A.M. The hours from noon until midnight are labeled P.M.

Helpful Hint
12:00 P.M. is noon.
12:00 A.M. is midnight.

Wake Up

9:00 A.M.

Look at the Moon

9:00 P.M.

Tell what time is shown for the activity. Write the time. Circle A.M or P.M.

1. Go to the Park

:   A.M.
    P.M.

2. Go Bowling

:   A.M.
    P.M.

3. Make your Bed

:   A.M.
    P.M.

Tell what time is shown for the activity.
Draw the hands on the clock. Circle A.M. or P.M.

4. Rooster Crowing

`6:00`

A.M.
P.M.

5. Flying a Kite

`2:30`

A.M.
P.M.

6. Going Shopping

`5:15`

A.M.
P.M.

7. Jay and Mike went to the county fair with Mike's parents. They got there at 10:00 A.M. They stayed for 4 and a half hours. Did they get home in the A.M or P.M.? _____

Hey, it's time to WAKE UP!

Vocabulary Check

Complete each sentence.

A.M.   **P.M.**

8. Casey has an art lesson at 4:30. _____ .

9. Sam eats dinner at 5:30 _____ .

10. Kaylee eats breakfast at 6:00 _____ .

Math at Home Several times in the next 24 hours, ask your child for the time and then if it is A.M. or P.M.

My Review

Vocabulary Check [Vocab abc]

Complete each sentence.

hour	half hour	quarter hour	minute
A.M.	P.M.	digital	analog

1. The long hand on an analog clock is the _____ hand.

2. The short hand on an analog clock is the _____ hand.

3. 15 minutes is a _____.

4. 30 minutes is a _____.

5. A clock that uses hands to show the time is an _____ clock.

6. A clock that uses numbers to show the time is a _____ clock.

7. The hours from noon to midnight are _____.

8. The hours from midnight to noon are _____.

Concept Check

Read the time. Write the time.

9.   :   10.   :   11.   :

Tell what time is shown. Draw the hands on each clock.

12.   7:35   13.   4:25   14.   6:55

Circle the best choice.

15. Walk the dog.   11:45   A.M.   P.M.

16. Go hiking.   2:30   A.M.   P.M.

17. Wash dinner dishes.   6:30   A.M.   P.M.

Name _____

Problem Solving

18. On a clock, the hour hand is between the 2 and 3. The minute hand is on the 5. What time is it? Show the time on both clocks.

Big Ben
It's a really big clock!

19. The time is 9:00. In 6 hours, Greg will leave school. What time will it be in 6 hours? Circle A.M. or P.M.

_____ A.M.   P.M.

20. Derek has baseball practice at 2:30. Luke has practice at 4:15. Show the times on the clocks. Write the time and circle if it would be A.M. or P.M.

Derek   Luke

_____ A.M.   _____ A.M.
        P.M.   P.M.

Test Practice

21. The boat picked up passengers every half hour. If the first boat arrives at 8:00 A.M., when will the fourth boat arrive?

12:00 P.M.   10:00 A.M.   9:30 A.M.   10:30 A.M.
   ○   ○   ○   ○

Show the ways to write and use time.

Draw the hands to
show the time.

4:00 Digital

Analog

Write the time.

Skip count by
5s. Tell the time.

Skip count by fives.
The time is _____.

**ESSENTIAL
QUESTION**

How do I use and
tell time?

Draw the hands.
Write the time.

quarter past 5   half past 5

You'll sail through this!

Chapter 11

Customary and Metric Lengths

ESSENTIAL QUESTION
How can I measure objects?

I Love Sports!

Watch a video!

1st

12 13 20

Watch

635

My Common Core State Standards CCSS

Measurement and Data

2.MD.1 Measure the length of an object by selecting and using appropriate tools such as rulers, yardsticks, meter sticks, and measuring tapes.

2.MD.2 Measure the length of an object twice, using length units of different lengths for the two measurements; describe how the two measurements relate to the size of the unit chosen.

2.MD.3 Estimate lengths using units of inches, feet, centimeters, and meters.

2.MD.4 Measure to determine how much longer one object is than another, expressing the length difference in terms of a standard length unit.

2.MD.5 Use addition and subtraction within 100 to solve word problems involving lengths that are given in the same units, e.g., by using drawings (such as drawings of rulers) and equations with a symbol for the unknown number to represent the problem.

2.MD.6 Represent whole numbers as lengths from 0 on a number line diagram with equally spaced points corresponding to the numbers 0, 1, 2, …, and represent whole-number sums and differences within 100 on a number line diagram.

2.MD.9 Generate measurement data by measuring lengths of several objects to the nearest whole unit, or by making repeated measurements of the same object. Show the measurements by making a line plot, where the horizontal scale is marked off in whole-number units.

Standards for Mathematical PRACTICE ▼

1. Make sense of problems and persevere in solving them.
2. Reason abstractly and quantitatively.
3. Construct viable arguments and critique the reasoning of others.
4. Model with mathematics.
5. Use appropriate tools strategically.
6. Attend to precision.
7. Look for and make use of structure.
8. Look for and express regularity in repeated reasoning.

= focused on in this chapter

Name _____

Am I Ready?

Check ✓ ← Go online to take the Readiness Quiz

Add or subtract.

1. $\begin{array}{r} 34 \\ +18 \\ \hline \end{array}$

2. $\begin{array}{r} 26 \\ -17 \\ \hline \end{array}$

3. $\begin{array}{r} 38 \\ +36 \\ \hline \end{array}$

Measure the length of the object in cubes.

4.

_____ cubes

5.

_____ cubes

6. Neil's hand is 5 cubes long.
    He measures this checkerboard.
    It is 3 hands long. How many
    cubes long is the checkerboard?

    _____ cubes

How Did I Do? Shade the boxes to show the problems you answered correctly.

1	2	3	4	5	6

My Math Words

Vocab
abc

Review Vocabulary

compare   longest   shortest

Find three classroom objects. List them. Compare their lengths. Write the longest object. Write the shortest object.

_____   _____   _____

Longest   **Shortest**

How did you compare your objects?

_____

_____

_____

Lesson 11–7

centimeter (cm)

7 cm

0 1 2 3 4 5 6 7
centimeters

Lesson 11–1

estimate

about 12 inches

Lesson 11–2

foot (ft)

1 foot

Lesson 11–1

inch (in)

2 in.

0 1 2 3
inches

Lesson 11–1

length

Lesson 11–1

measure

0 1 2
inches

Directions:
Ideas for Use

• Have students group 2 or 3 common words. Ask them to add a word that is unrelated to the group. Have them ask a friend to name the unrelated word.

• Ask students to find pictures to show an example of each word. Have them ask a friend to guess which word the picture shows.

To find a number close to an exact amount.

A metric unit for measuring length.

A customary unit for measuring length.
The plural is inches.
12 inches = 1 foot

A customary unit for measuring length.
The plural is feet.
1 foot = 12 inches

To find length using standard or nonstandard units.

How long or how far away something is.

My Vocabulary Cards

Vocab abc

Mathematical PRACTICE

Lesson 11-7

meter (m)

15 meters

Lesson 11-2

yard (yd)

1 yard

Directions:
More Ideas for Use

- Have students use a blank card to write this chapter's essential question. Have them use the back of the card to write or draw examples that help them answer the question.

- Ask students to use the blank cards to write some of the units of measure they learned in this chapter. They should give examples of different objects they would measure with each unit on the back of each card.

A customary unit for measuring length.
I yard = 3 feet or 36 inches

A metric unit used to measure length.
I meter = 100 centimeters

My Foldable

FOLDABLES® Follow the steps on the back to make your Foldable.

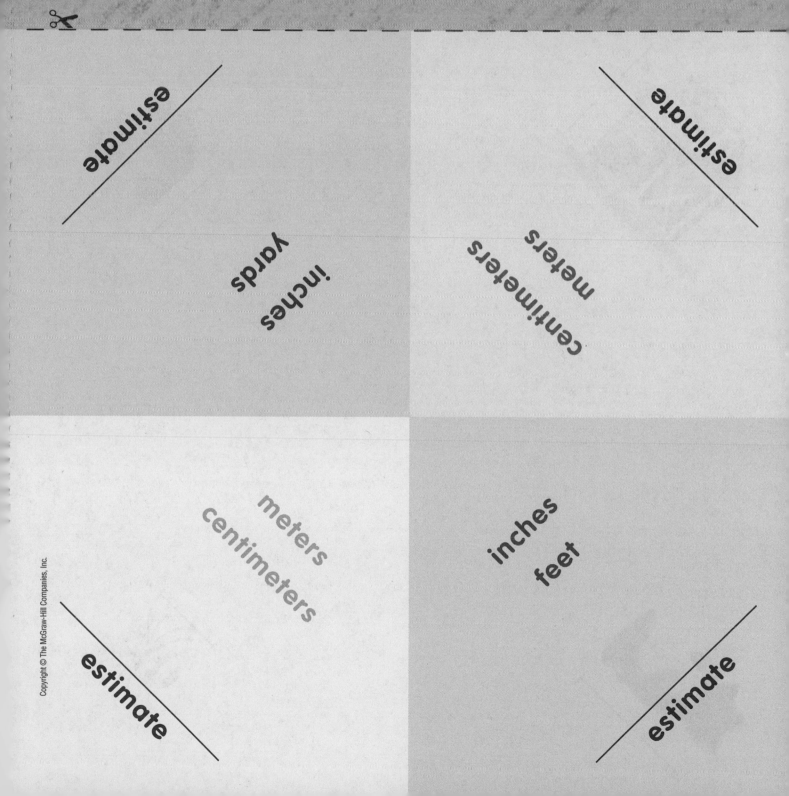

estimate

yards

inches

estimate

centimeters

meters

meters
centimeters

inches
feet

estimate

estimate

Name

Inches

Lesson 1

ESSENTIAL QUESTION
How can I measure objects?

Explore and Explain Watch Tools

Hold still up there!

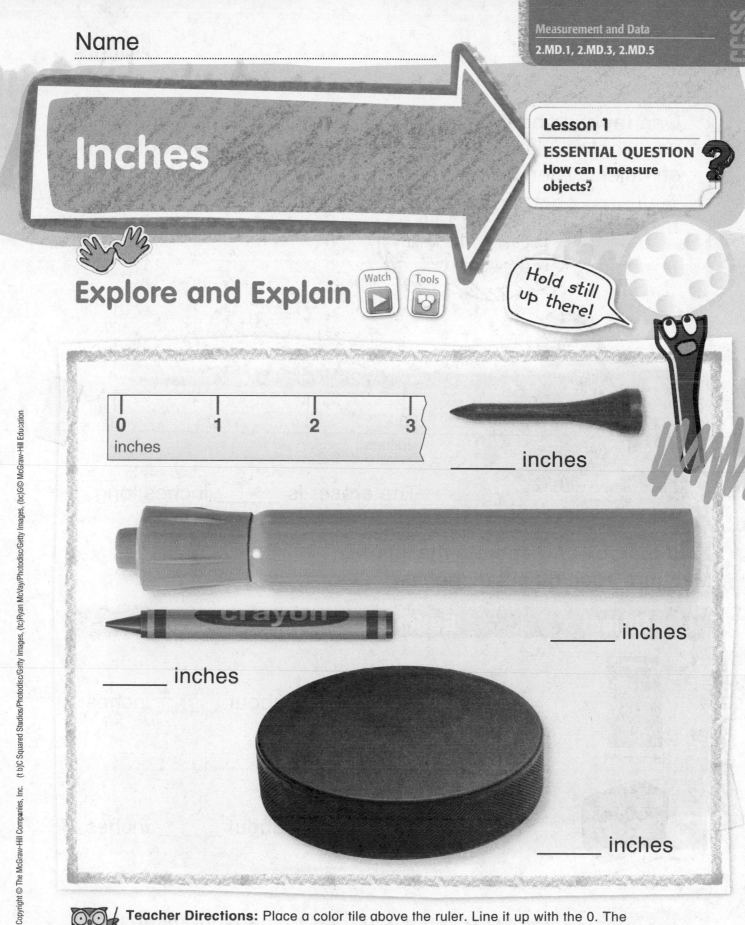

```
0       1       2       3
inches
```
_____ inches

_____ inches

_____ inches

_____ inches

Teacher Directions: Place a color tile above the ruler. Line it up with the 0. The color tile is one inch long. Use color tiles to measure the length of each object on the page. Write each length in inches.

See and Show

One **inch** is about the **length** of one color tile. You can use what you know about inches to **estimate** the length of an object. Then use an inch ruler to **measure** the length.

Helpful Hint
Estimate: The length is about 3 inches.

The eraser is _____ inches long.

Find the object. Estimate the length.
Measure each object in inches.

Object	Estimate	Measure
1. CHALK	about _____ inches	about _____ inches
2.	about _____ inches	about _____ inches

Talk Math

How do you use a ruler to measure inches?

On My Own

**Find the object. Estimate the length.
Measure each object in inches.**

Object	Estimate	Measure
3.	about _____ inches	about _____ inches
4.	about _____ inches	about _____ inches
5.	about _____ inches	about _____ inches
6.	about _____ inches	about _____ inches
7.	about _____ inches	about _____ inches

8. Kaya's surfboard must be at least 15 inches longer than 48 inches. Should Kaya choose a surfboard that is 55 inches long or 65 inches long?

_____ inches long

9. Owen's foot is 6 inches long. His dad's foot is 6 inches longer than Owen's foot. How long is his dad's foot?

_____ inches

10. Carter estimates his math book is 7 inches long. Wyatt estimates it is 15 inches long. The book is 8 inches long. Whose estimate is closer?

_____

Write Math How are an estimate and an actual measurement different?

_____

_____

_____

Name ....

My Homework

Homework Helper

eHelp   Need help? connectED.mcgraw-hill.com

Estimate the length of an object. Then check your estimate by measuring the length with an inch ruler.

Helpful Hint
A small paper clip is about 1 inch long.

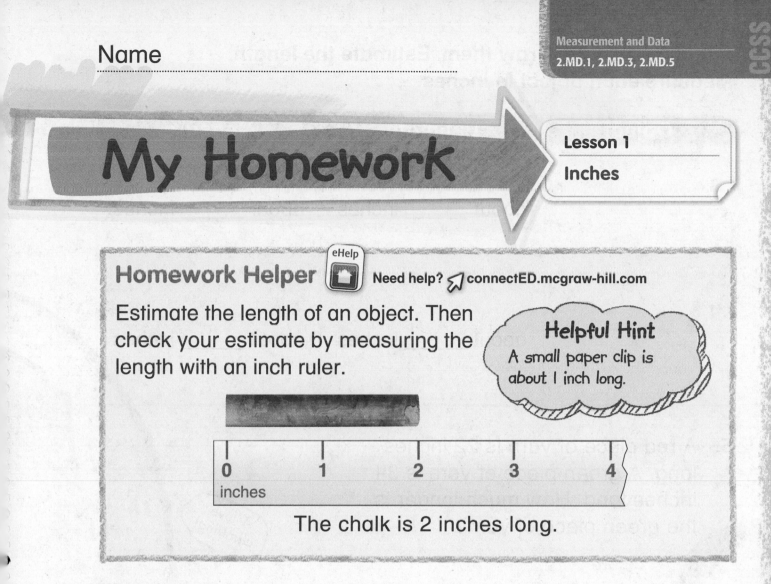

```
0        1        2        3        4
inches
```

The chalk is 2 inches long.

Practice

**Find the object. Estimate the length.
Measure each object in inches.**

Object	Estimate	Measure
I.	about _____ inches	about _____ inches
2.	about _____ inches	about _____ inches

Find two objects. Draw them. Estimate the length. Measure each object in inches.

Object	Estimate	Measure
3.	about _____ inches	about _____ inches
4.	about _____ inches	about _____ inches

5. A red piece of yarn is 22 inches long. A green piece of yarn is 34 inches long. How much longer is the green piece of yarn?

_____ inches

I love climbing!

Vocabulary Check

Circle the correct answer.

estimate   inch   measure   length

6. An _____ is a unit for measuring length.

🦉 **Math at Home** Ask your child to measure the length of a fork and a spoon using an inch ruler.

Feet and Yards

Explore and Explain

Teacher Directions: Discuss each object in the picture. Circle the objects that could be measured with an inch ruler. Draw an X on the objects that are too big to be measured with an inch ruler.

Online Content at connectED.mcgraw-hill.com

See and Show

You can measure in feet or yards. A **foot** is equal to 12 inches. A **yard** is equal to 36 inches. Length can be measured in any direction.

Helpful Hint
3 rulers is 3 feet.

3 rulers

Helpful Hint
1 yardstick is 1 yard.

1 yardstick

about ___3___ feet

about ___1___ yard

Find the object. Estimate the length.
Measure each object in feet or yards.

Object	Estimate	Measure
1.	about _____ feet	about _____ feet
2.	about _____ yards	about _____ yards

Talk Math How can you measure a large object with a ruler?

Name _____

On My Own

Find the object. Estimate the length. Measure each object in feet or yards.

Object	Estimate	Measure
3.	about _____ feet	about _____ feet
4.	about _____ feet	about _____ feet
5.	about _____ feet	about _____ feet
6.	about _____ feet	about _____ feet
7.	about _____ yard	about _____ yard

Problem Solving

8. The length of Sydney's tennis racket is 2 rulers long. How many feet long is Sydney's tennis racket?

_____ feet

9. Gabriella is 2 inches taller than a yard. How many inches tall is Gabriella?

Sydney has an awesome serve!

_____ inches

10. The pool at the park is 18 feet wide. How many yardsticks would it take to measure the width of the pool?

_____ yardsticks

HOT Problem Josh used 10 yardsticks to measure the length of his basement. How many rulers would he have used? Explain.

_____

_____

_____

My Homework

Homework Helper

eHelp

Need help? connectED.mcgraw-hill.com

One foot is equal to 12 inches. One yard is equal to 36 inches. A ruler is 1 foot long. A yardstick is 1 yard long.

Helpful Hint
3 feet = 36 inches

ruler

yardstick

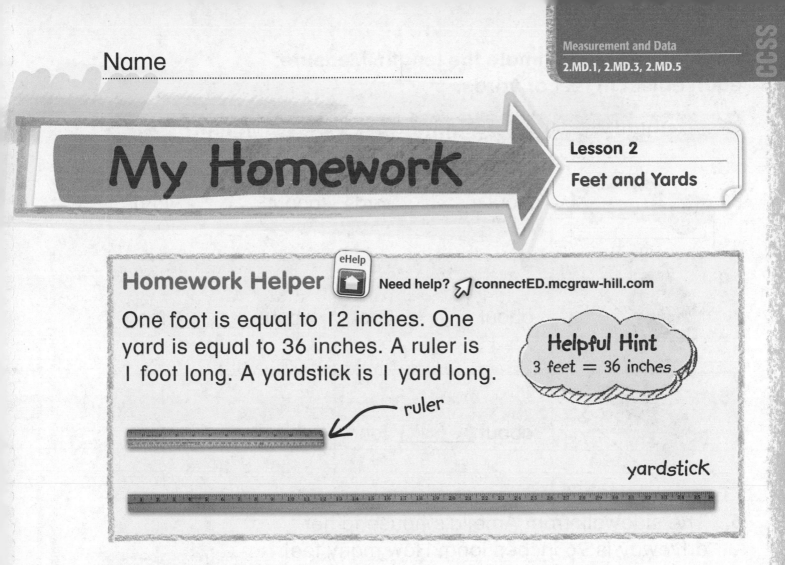

Practice

Find the object. Estimate the length. Measure each object in feet or yards.

Object	Estimate	Measure
1.	about _____ feet	about _____ feet
2.	about _____ yards	about _____ yards

Find the object. Estimate the length. Measure each object in feet or yards.

Object	Estimate	Measure
3.	about _____ yards	about _____ yards
4.	about _____ foot	about _____ foot
5.	about _____ feet	about _____ feet

6. The sidewalk from Ameila's house to her driveway is 96 inches long. How many feet long is the sidewalk?

_____ feet

Sidewalk race!

Hey, wait up!

Vocabulary Check

Circle the correct answer.

7. **foot**

    3 yards   12 inches   36 inches   1 yard

Math at Home Ask your child to measure his or her room using a yardstick or a ruler.

Name

Select and Use Customary Tools

Copyright © The McGraw-Hill Companies, Inc. (l)Ingram Publishing/Alamy Images, (r)Brand X Pictures/Getty Images

Lesson 3

ESSENTIAL QUESTION
How can I measure objects?

I love rulers because I'm an inch worm!

Explore and Explain

1.

2.

3.

Teacher Directions: Draw an object you would measure with an inch ruler in box 1. Draw an object you would measure with a yardstick in box 2. Draw an object you would measure with a measuring tape in box 3. Explain your drawings.

See and Show

You can select and use tools to measure length. Circle the tool you would use to measure a marker.

Helpful Hint
Measure objects shorter than a foot with an inch ruler, objects longer than a foot with a yardstick, and objects longer than 3 feet with a measuring tape.

inch ruler

yardstick

marker

measuring tape

**Find the object. Choose the tool and measure it.
Explain why you chose that tool.**

Object	Tool	Measure
1.	_____	about _____
2.	_____	about _____
3.	_____	about _____

Talk Math How do you know which tool to use to measure?

Name _____

On My Own

**Find the object. Choose the tool and measure it.
Explain why you chose that tool.**

Object	Tool	Measure
4.	_____	about _____
5.	_____	about _____
6.	_____	about _____
7.	_____	about _____
8.	_____	about _____
9.	_____	about _____

Problem Solving

10. Vijay measures the length of his bike with a yardstick. He says it is one yardstick long. How many inches long is it?

_____ inches

Boy, I'm sweaty after gym class!

11. Ethan measured his foot with a yardstick. What other tool would have been a better choice?

_____

12. Aisha measured her backyard with a ruler. Curtis measured his backyard with a yardstick. What other tool would have been a better choice?

_____

HOT Problem Omar laid 6 yardsticks end to end to measure his driveway. He said it was 6 feet long. Tell why Omar is wrong. Make it right.

_____

_____

_____

Name _____

My Homework

Lesson 3

Select and Use Customary Tools

Homework Helper

Need help? connectED.mcgraw-hill.com

You can select and use tools to measure length.

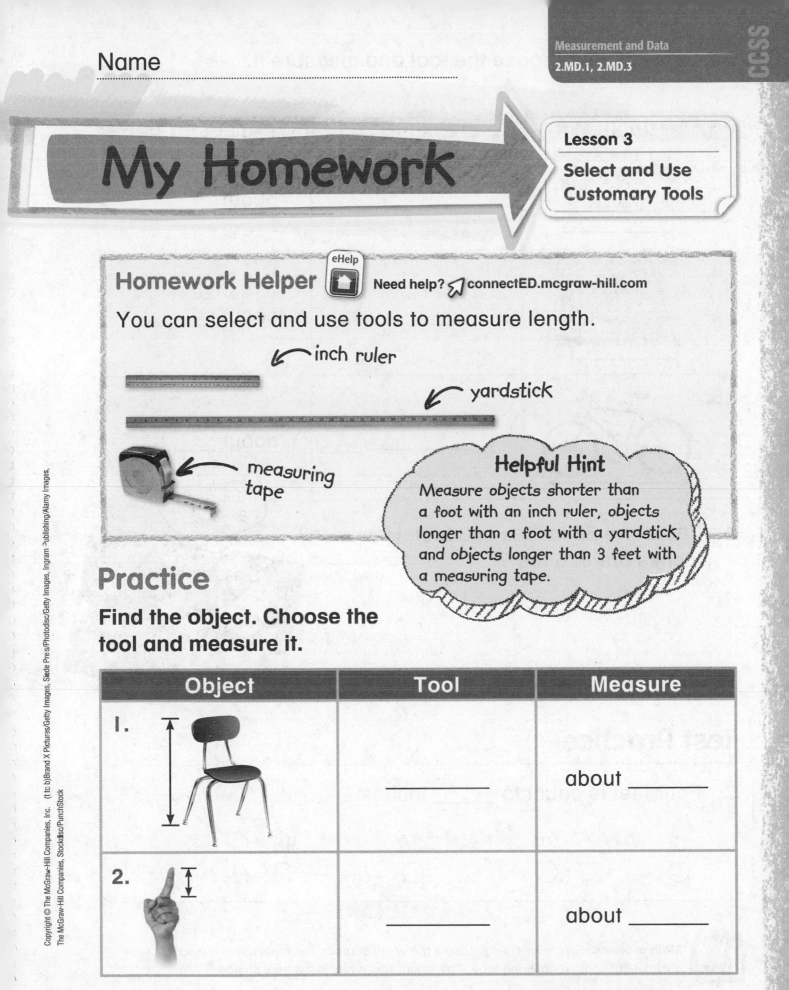

← inch ruler

← yardstick

← measuring tape

Helpful Hint

Measure objects shorter than a foot with an inch ruler, objects longer than a foot with a yardstick, and objects longer than 3 feet with a measuring tape.

Practice

Find the object. Choose the tool and measure it.

Object	Tool	Measure
1.	_____	about _____
2.	_____	about _____

Find the object. Choose the tool and measure it.

Object	Tool	Measure
3. 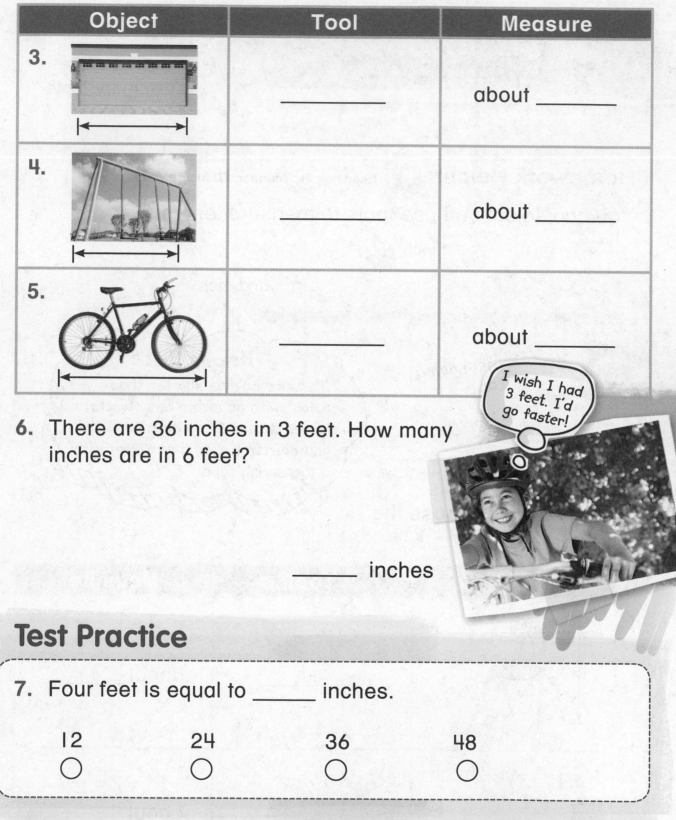	_____	about _____
4.	_____	about _____
5.	_____	about _____

6. There are 36 inches in 3 feet. How many inches are in 6 feet?

> I wish I had 3 feet. I'd go faster!

_____ inches

Test Practice

7. Four feet is equal to _____ inches.

   12   24   36   48
   ○   ○   ○   ○

Math at Home Have your child measure the width of his or her room using a ruler, a yardstick, and a measuring tape. Talk about which tool is the best choice.

Name _____

Check My Progress

Vocabulary Check Vocab

Draw a line to match each word to the correct sentence.

1. **length**   To find a number close to an exact amount.

2. **inch**   Distance from end to end.

3. **estimate**   A customary unit for measuring length.

Concept Check

Find the object. Estimate the length.
Measure each object.

Object	Estimate	Measure
4.	about _____ inches	about _____ inches
5.	about _____ feet	about _____ feet
6.	about _____ yards	about _____ yards

Find the object. Choose the tool and measure it.

Object	Tool	Measure
7.	_____	about _____
8.	_____	about _____
9.	_____	about _____

10. James measured the fence in his backyard with a ruler. It was 24 feet long. Which tool should James have used to measure the fence?

_____

Test Practice

11. Mr. Tom's driveway measures 7 yards long. How many feet long is Mr. Tom's driveway?

7   14   21   28

○   ○   ○   ○

Name

Compare Customary Lengths

Lesson 4

ESSENTIAL QUESTION
How can I measure objects?

Explore and Explain

I'm taller!

Helpful Hint
Measure length up and down to measure how tall.

Teacher Directions: Find two objects in your classroom. Compare the lengths. Draw the objects. Explain which object is shorter and which object is longer.

See and Show

You can compare lengths of objects.

baseball bat

golf club

_____ 33 _____ inches

_____ 31 _____ inches

The baseball bat is _____ 33 _____ inches long.

The golf club is _____ 31 _____ inches long.

The golf club is _____ 2 _____ inches shorter than the bat.

Think:

33
−31
‾‾‾‾
2

**Find the objects. Measure them. Write the lengths.
Write longer or shorter.**

1.

_____ inches   _____ inches

The pencil is _____ inches _____ .

2.

_____ feet   _____ feet

The desk is _____ feet _____ .

Talk Math Why do you need to know how
to compare lengths?

Name _____

On My Own

Find the objects. Measure them. Write the lengths.
Write longer or shorter.

3.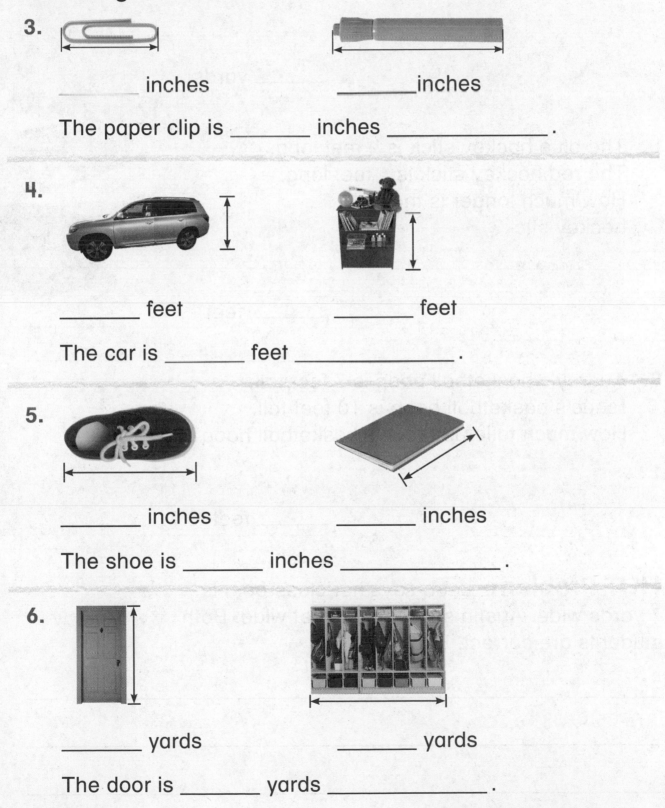

_____ inches   _____ inches

The paper clip is _____ inches _____.

4.

_____ feet   _____ feet

The car is _____ feet _____.

5.

_____ inches   _____ inches

The shoe is _____ inches _____.

6.

_____ yards   _____ yards

The door is _____ yards _____.

Problem Solving

7. The playground is 90 yards long. The field by my house is 80 yards long. How much longer is the playground?

_____ yards

8. The blue hockey stick is 4 feet long. The red hockey stick is 6 feet long. How much longer is the red hockey stick?

_____ feet

9. Audrey's basketball hoop is 7 feet tall. Isaac's basketball hoop is 10 feet tall. How much taller is Isaac's basketball hoop?

_____ feet

HOT Problem Leah says her classroom is 7 yards wide. Austin says it is 21 feet wide. Both students are correct. Explain.

_____

_____

_____

Name _____

My Homework

Homework Helper

eHelp   Need help? connectED.mcgraw-hill.com

You can compare the lengths of objects.

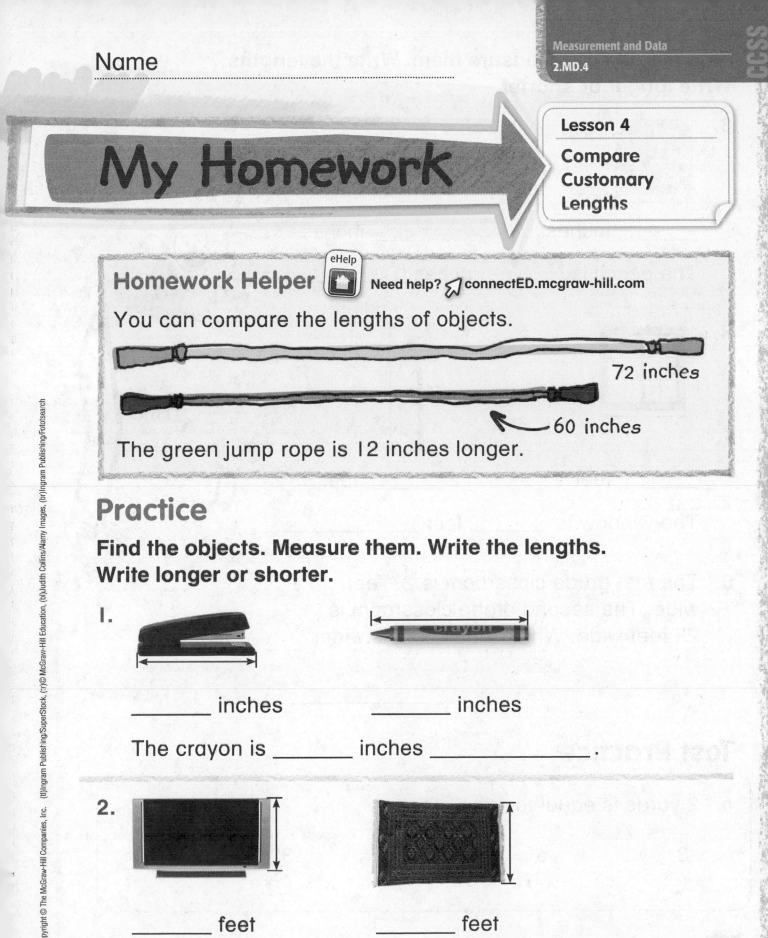

72 inches

60 inches

The green jump rope is 12 inches longer.

Practice

Find the objects. Measure them. Write the lengths.
Write longer or shorter.

1.

_____ inches   _____ inches

The crayon is _____ inches _____.

2.

_____ feet   _____ feet

The rug is _____ feet _____.

**Find the objects. Measure them. Write the lengths.
Write longer or shorter.**

3.

_____ inches   _____ inches

The pencil is _____ inches _____.

I'm a birdie!

4.

_____ feet   _____ feet

The window is _____ feet _____.

Hi!

5. The first grade classroom is 27 feet wide. The second grade classroom is 24 feet wide. Which classroom is wider?

_____

Test Practice

6. 2 yards is equal to _____ feet.

3 ◯   6 ◯   24 ◯   32 ◯

Math at Home Have your child measure two rooms in your house and compare the lengths.

Name _____

Relate Inches, Feet, and Yards

Lesson 5

ESSENTIAL QUESTION
How can I measure objects?

Explore and Explain [Watch ▶]

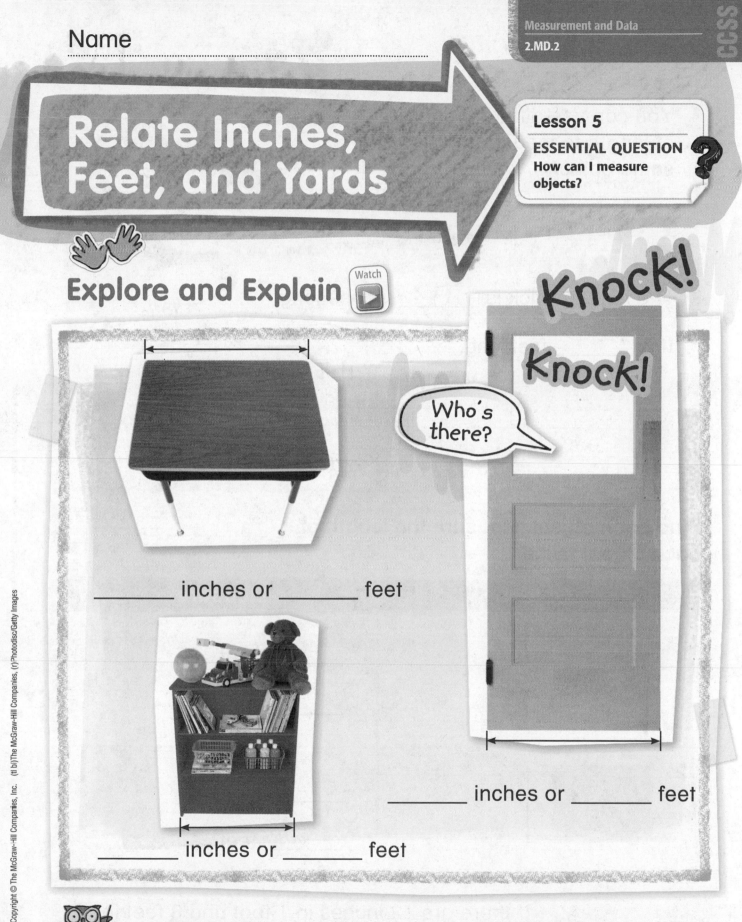

_____ inches or _____ feet

_____ inches or _____ feet

_____ inches or _____ feet

Knock!

Knock!

Who's there?

Teacher Directions: Find each object in the classroom. Measure each object in inches. Measure each object in feet. Discuss the measurements.

Online Content at 🖝 **connectED.mcgraw-hill.com**

Chapter 11 • Lesson 5   671

See and Show

You can use different units of length to measure the same object.

The hockey stick is __48__ inches long.

It is __4__ feet long.

48 inches

Helpful Hint
12 inches = 1 foot
3 feet = 1 yard

Find each object. Measure the length of each object twice.

Object	Measure
1.	_____ feet _____ yards
2.	_____ yards _____ feet

Talk Math If there are 12 inches in 1 foot and 3 feet in 1 yard, how many inches are in 1 yard?

Name _____

On My Own

Find each object. Measure the length of each object twice.

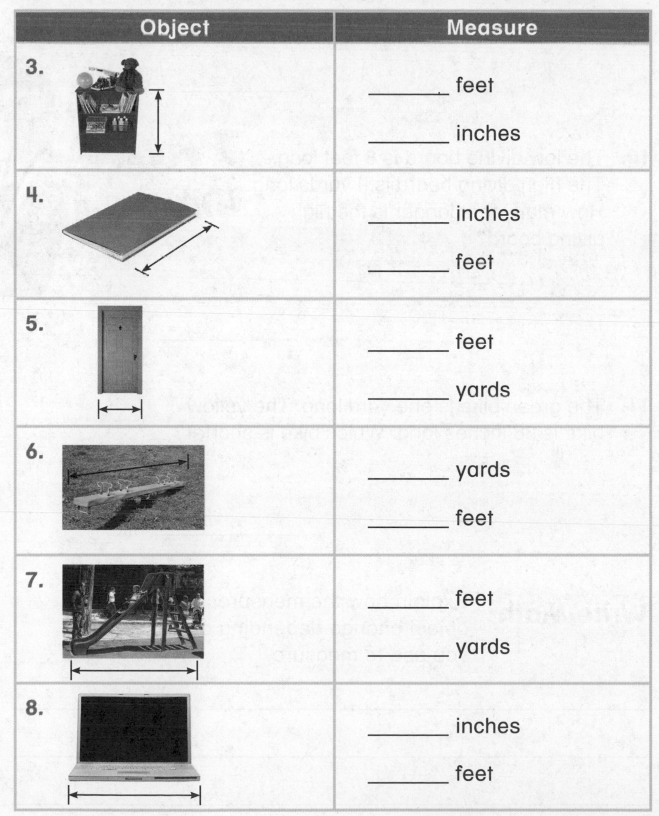

Object	Measure
3.	_____ feet _____ inches
4.	_____ inches _____ feet
5.	_____ feet _____ yards
6.	_____ yards _____ feet
7.	_____ feet _____ yards
8.	_____ inches _____ feet

Problem Solving

9. Sheradon's scooter is 3 feet long. Sage's scooter is 38 inches long. Whose scooter is longer?

_____

10. The low diving board is 8 feet long. The high diving board is 4 yards long. How many feet longer is the high diving board?

I rule the pool!

_____ feet

11. The green bike is one yard long. The yellow bike is 38 inches long. Which bike is shorter?

_____ bike

Write Math   Explain how the measurements of an object change depending on which unit you use to measure.

_____

_____

_____

My Homework

Homework Helper 🏠 eHelp   Need help? 👉 connectED.mcgraw-hill.com

You can use different units of length to measure the same object.

The rope is 21 feet long.

It is 7 yards long.

21 feet or 7 yards →

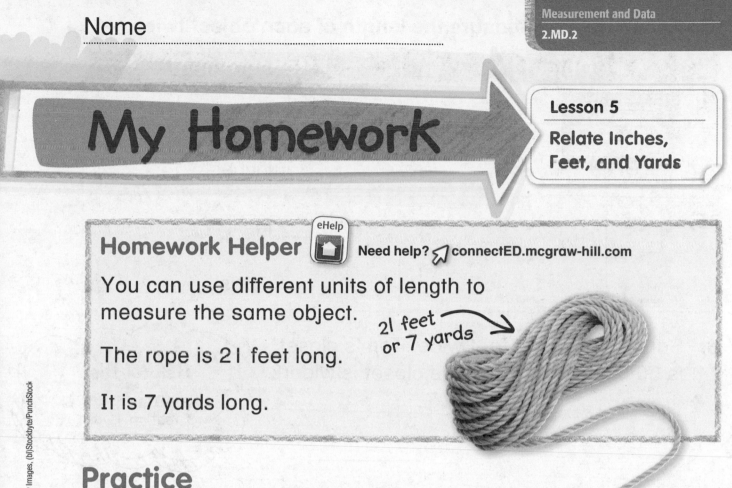

Practice

Find each object. Measure the length of each object twice.

Object	Measurement
1.	_____ feet _____ yards
2.	_____ yards _____ feet
3.	_____ inches _____ feet

Find each object. Measure the length of each object twice.

Object	Measurement
4.	_____ feet _____ inches
5.	_____ feet _____ yards

6. Sam's closet is 4 feet wide. Tom's closet is 50 inches wide. Whose closet is wider?

Helpful Hint
12 inches = 1 foot
3 feet = 1 yard

_____

7. The fence in Laura's backyard is 15 feet long. How many yards long is the fence?

_____ yards

Test Practice

8. 72 inches is equal to _____ yards.

   1   2   3   4
   ○   ○   ○   ○

Math at Home Have your child measure an object in your house using inches, feet, and yards. Discuss how each measurement relates to the size of the unit.

Name _____

Problem Solving

STRATEGY: Use Logical Reasoning

Koko wants to plant a garden. She cannot decide if it should be 10 inches, 10 feet, or 100 yards long. About how long should the garden be?

1 Understand

Underline what you know.
Circle what you need to find.

2 Plan

How will I solve the problem?

3 Solve

Use logical reasoning.

__10__ inches is too short.

__100__ yards is too long.

The garden should be __10__ feet long.

4 Check

Is my answer reasonable? Explain.

Practice the Strategy

Zachary is looking up at the high diving board at the pool. Is the high diving board 10 inches high, 10 feet high, or 10 yards high?

That's my best cannonball EVER!

1 Understand <u>Underline</u> what you know.
(Circle) what you need to find.

2 Plan How will I solve the problem?

3 Solve I will...

4 Check Is my answer reasonable? Explain.

Apply the Strategy

1. Sam planted a tomato plant that is
   1 foot tall. The plant grows a little
   each week. After 4 weeks, would the
   plant be 10 inches or 14 inches tall?

_____ inches

2. Jane made a paper chain that is 1 yard long.
   Brad made a paper chain that is 2 feet long.
   Who made the longer paper chain?

_____

3. Mr. Moore's class is collecting things
   to measure. Lisa finds a pinecone.
   Would the pinecone be 3 inches,
   3 feet, or 3 yards long?

_____

Review the Strategies

Choose a strategy
- Use logical reasoning.
- Write a number sentence.
- Make a model.

4. Santos connected 6 markers. Each marker measures 5 inches. How long are all the markers?

_____ inches

5. Jamil had a pencil that was 6 inches long. After a month it was an inch long. How many inches did Jamil use?

I may be short, but I get a lot done!

_____ inches

6. Suni measured from one end of her kitchen table to the middle. It was 42 inches long. Then she measured from the middle to the other end. It was the same length. How long is the table?

_____ inches

Name _____

My Homework

eHelp

Dave is more than 40 inches tall. He is less than 43 inches tall. His height is an even number. How tall is Dave?

Dad says I'm growing like a weed!

1 Understand Underline what you know.
Circle what you need to find.

2 Plan How will I solve the problem?

3 Solve Use logical reasoning.
Dave is either 41 or 42 inches tall.
42 is an even number.
Dave is 42 inches tall.

4 Check Is my answer reasonable?

Problem Solving

Underline what you know. Circle what you need to find. Use logical reasoning to solve.

1. A bookshelf is 14 inches wide. There is enough space to fit 2 trophies. One trophy is 2 inches wider than the other. How wide is the wider trophy?

    _____ inches

2. Joan has a stalk of celery that is 6 inches long. She cuts it into two pieces. Each piece is the same length. How long is each piece.

    _____ inches

3. Marlon and Brent are each measuring one of their shoes. Their shoes measure 11 inches in all. Marlon's shoe is 1 inch longer than Brent's shoe. How long is Brent's shoe?

    _____ inches

Test Practice

4. 36 inches is _____ feet.

    1 ○   2 ○   3 ○   4 ○

Math at Home Ask your child to estimate distances from one room to another. Have our child check by measuring.

Check My Progress

Vocabulary Check

Draw lines to match.

1. **measure**   To find the length using standard or nonstandard units.

2. **foot**   Equal to 36 inches.

3. **yard**   Equal to 12 inches.

Concept Check

Find the objects. Measure them. Write the lengths. Write longer or shorter.

4.

_____ inches   _____ inches

The marker is _____ inches _____.

5.

_____ feet   _____ feet

The cabinet is _____ feet _____.

Find each object. Measure the length of each object twice.

Object	Measure
6.	_____ feet _____ yards
7.	_____ inches _____ feet
8.	_____ inches _____ feet

9. Bill measured his favorite book with a yardstick. It was 10 inches long. Which tool should Bill have used to measure the book?

_____

Test Practice

10. Emily drew a blue line that is 12 inches long. She drew a purple line that is 4 inches long. How long are both lines in all?

12 inches   14 inches   16 inches   124 inches

○   ○   ○   ○

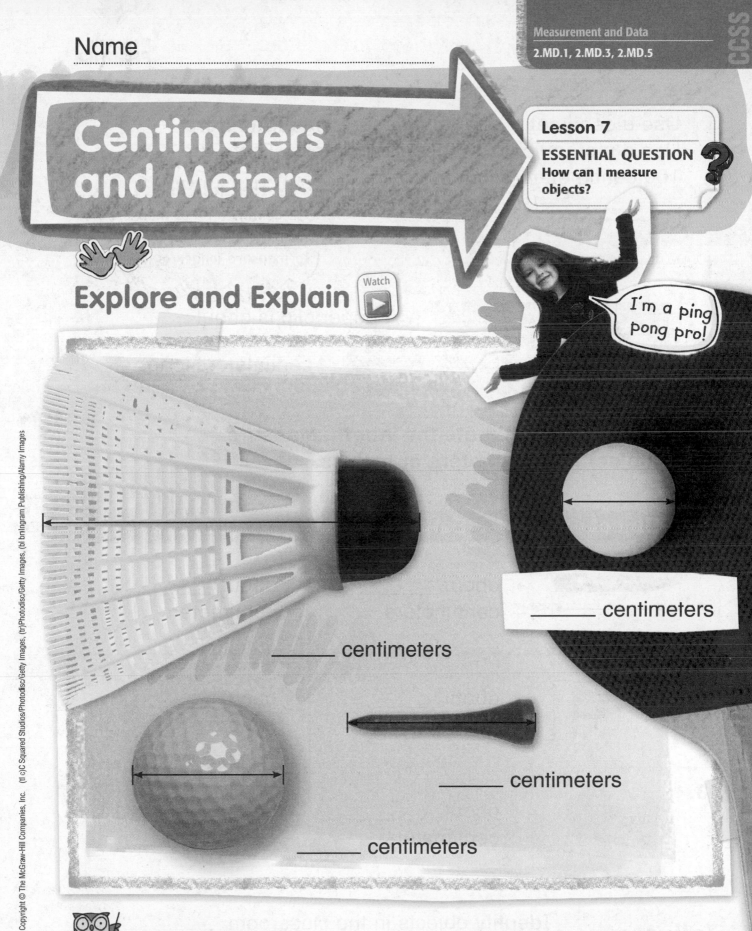

Name

Centimeters and Meters

Lesson 7

ESSENTIAL QUESTION
How can I measure objects?

Explore and Explain Watch ▶

I'm a ping pong pro!

_____ centimeters

_____ centimeters

_____ centimeters

_____ centimeters

Teacher Directions: One unit cube is I centimeter long. Use unit cubes to measure each object. Write the length of each object on the line.

See and Show

Use a centimeter ruler to measure in **centimeters**.
Use a meterstick to measure in **meters**. There are
100 centimeters in a meter.

Helpful Hint
Use centimeters to measure
shorter objects. Use meters
to measure longer objects.

0 1 2 3 4 5
centimeters

A paper clip is about
__5__ centimeters long.

**Find the object. Estimate the length. Measure
each object in centimeters or meters.**

Object	Estimate	Measure
1.	about _____ centimeters	about _____ centimeters
2.	about _____ meters	about _____ meters
3.	about _____ centimeters	about _____ centimeters

Talk Math Identify objects in the classroom
that are about 1 centimeter long.

Name

On My Own

Find the object. Estimate the length. Measure each object in centimeters or meters.

Object	Estimate	Measure
4.	about _____ centimeters	about _____ centimeters
5.	about _____ centimeters	about _____ centimeters
6.	about _____ centimeters	about _____ centimeters
7.	about _____ meters	about _____ meters
8.	about _____ meters	about _____ meters

Problem Solving

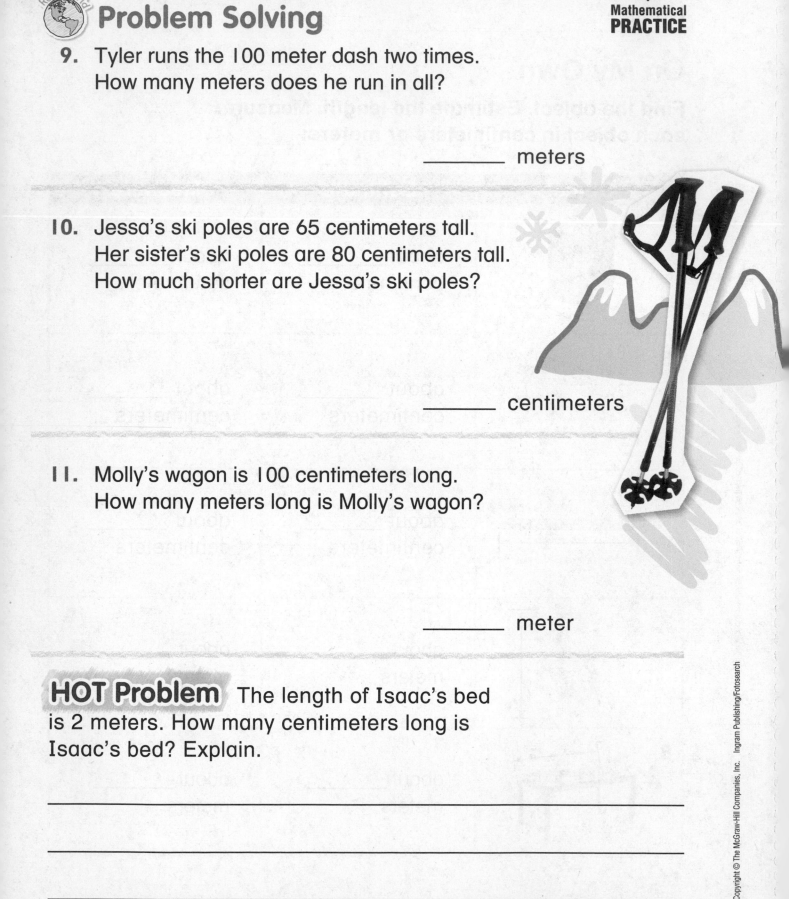

Mathematical PRACTICE

9. Tyler runs the 100 meter dash two times. How many meters does he run in all?

_____ meters

10. Jessa's ski poles are 65 centimeters tall. Her sister's ski poles are 80 centimeters tall. How much shorter are Jessa's ski poles?

_____ centimeters

11. Molly's wagon is 100 centimeters long. How many meters long is Molly's wagon?

_____ meter

HOT Problem The length of Isaac's bed is 2 meters. How many centimeters long is Isaac's bed? Explain.

_____

_____

_____

Name

My Homework

Homework Helper

eHelp

Need help? connectED.mcgraw-hill.com

The cube is 1 centimeter long. There are 100 centimeters in 1 meter.

centimeters

Practice

Find the object. Estimate the length. Measure each object in centimeters or meters.

Object	Estimate	Measure
1.	about _____ centimeters	about _____ centimeters
2.	about _____ centimeters	about _____ centimeters
3.	about _____ meters	about _____ meters

Find the object. Estimate the width. Measure each object in centimeters or meters.

Object	Estimate	Measure
4.	about _____ centimeters	about _____ centimeters
5.	about _____ meters	about _____ meters

6. One pine tree is 4 meters tall. The other pine tree is 3 meters tall. How many centimeters taller is the first pine tree?

_____ centimeters

Vocabulary Check

7. Circle the answer that is the same as 1 **meter**.

1 centimeter   10 centimeters

100 centimeters   1,000 centimeters

Math at Home Have your child identify objects he or she would measure using centimeters.

Name

Select and Use Metric Tools

Lesson 8

ESSENTIAL QUESTION
How can I measure objects?

I will now perform a half-twist!

Explore and Explain ▶Watch

1.

2.

Teacher Directions: Draw a picture of an object you would measure with a centimeter ruler in box 1. Draw a picture of an object you would measure with a meterstick in box 2.

See and Show

A centimeter ruler measures smaller objects.
A meterstick measures larger objects.
Circle the tool you would use to measure
a lunchbox.

centimeter ruler

meterstick

**Find the object. Choose the tool and measure it.
Explain why you chose that tool.**

Object	Tool	Measure
1.	_____	about _____
2.	_____	about _____
3.	_____	about _____

Talk Math   Can you measure a paper clip
with a meterstick? Explain.

Name

On My Own

Find the object. Choose the tool and measure it.
Explain why you chose that tool.

Object	Tool	Measure
4.	_____	about _____
5.	_____	about _____
6.	_____	about _____
7.	_____	about _____
8.	_____	about _____

Problem Solving

9. A wall in Jim's garage is 5 meters long. He is painting 2 meters of the wall blue. He wants to paint the rest of the wall red. How many meters will he paint red?

_____ meters

10. A sandbox is 300 centimeters long. How many meters long is it?

_____ meters

11. Allison swam 50 meters in the morning. She swam 40 meters in the evening. How many meters did Allison swim in all?

_____ meters

  Why do you need to know customary and metric length?

_____

_____

_____

Name _____

My Homework

Lesson 8
Select and Use Metric Tools

Homework Helper

Need help? connectED.mcgraw-hill.com

A centimeter ruler is used to measure smaller objects.

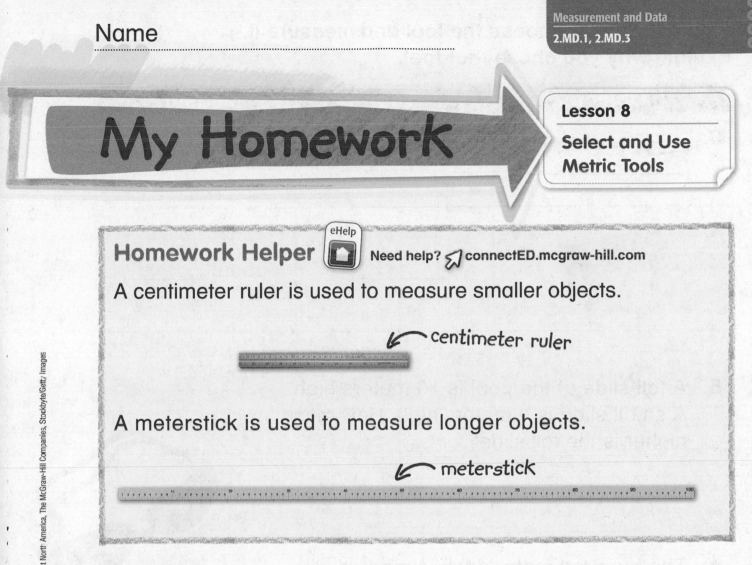

← centimeter ruler

A meterstick is used to measure longer objects.

← meterstick

Practice

Find the object. Choose the tool and measure it.
Explain why you chose that tool.

Object	Tool	Measure
1.	_____	about _____
2.	_____	about _____

Find the object. Choose the tool and measure it.
Explain why you chose that tool.

Object	Tool	Measure
3.	_____	about _____
4.	_____	about _____

5. A tall slide at the pool is 10 meters high.
   A small slide is 4 meters high. How much
   higher is the tall slide?

   _____ meters

6. The hundred meter dash is marked with
   a blue line. The 400 meter dash is marked
   with a red line. How much farther is the red line?

   _____ meters

Test Practice

7. Eight meters is equal to _____ centimeters.

   400   600   800   900
   ○   ○   ○   ○

 Math at Home Have your child identify objects he or she would measure
using meters and centimeters.

Copyright © The McGraw-Hill Companies, Inc. (t)Creatas/SuperStock, (c)Brand X Pictures/PunchStock, (br)Ingram Publishing/SuperStock

Name ....

Compare Metric Lengths

Lesson 9
ESSENTIAL QUESTION
How can I measure objects?

Explore and Explain

Measuring centimeters is a KICK!

_____ centimeters   _____ centimeters

🦉 **Teacher Directions:** Find two objects in your classroom. Draw a picture of each object. Measure each object in centimeters. Write how many centimeters long each object is. Circle the object that is longer.

Online Content at ⟋ **connectED.mcgraw-hill.com**   Chapter 11 • Lesson 9

See and Show

You can compare lengths of objects.

69 centimeters

76 centimeters

The blue tennis racket is ___69___ centimeters long.

The green tennis racket is ___76___ centimeters long.

The blue racket is ___7___ centimeters shorter than the green racket.

Find the objects. Measure them. Write the lengths.
Write longer or shorter.

1.

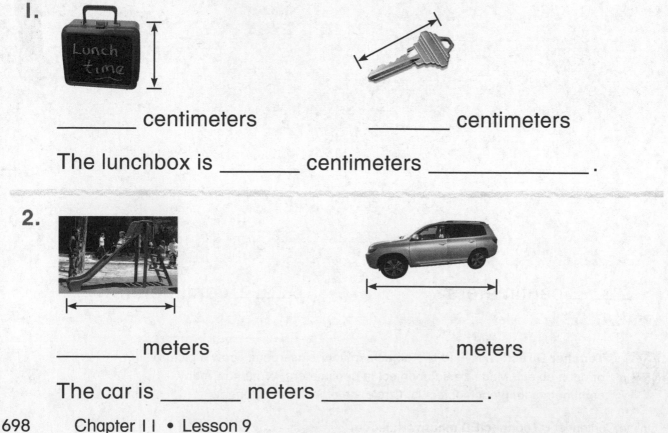

_____ centimeters   _____ centimeters

The lunchbox is _____ centimeters _____ .

2.

_____ meters   _____ meters

The car is _____ meters _____ .

Name _____

On My Own

Find the objects. Measure them. Write the lengths.
Write longer or shorter.

3.

_____ meters   _____ meters

The slide is _____ meters _____ .

4.

_____ centimeters   _____ centimeters

The knife is _____ centimeters _____ .

5.

_____ meters   _____ meters

The swing set is _____ meters _____ .

6.

_____ centimeters   _____ centimeters

The hair brush is _____ centimeters _____ .

Problem Solving

7. The red skateboard is 75 centimeters long. The purple skateboard is 83 centimeters long. How much longer is the purple skateboard?

Prepare for takeoff!

_____ centimeters

8. Marcus ran 50 meters. Will ran 67 meters. How much farther did Will run?

_____ meters

9. Alec rode his bike 82 meters. Kyla rode her bike 14 meters farther than Alec. How far did Kyla ride her bike?

_____ meters

Write Math How can you compare a measurement in centimeters to a measurement in meters?

_____

_____

My Homework

Lesson 9

Compare Metric Lengths

Homework Helper

eHelp

Need help? connectED.mcgraw-hill.com

You can compare lengths of objects.

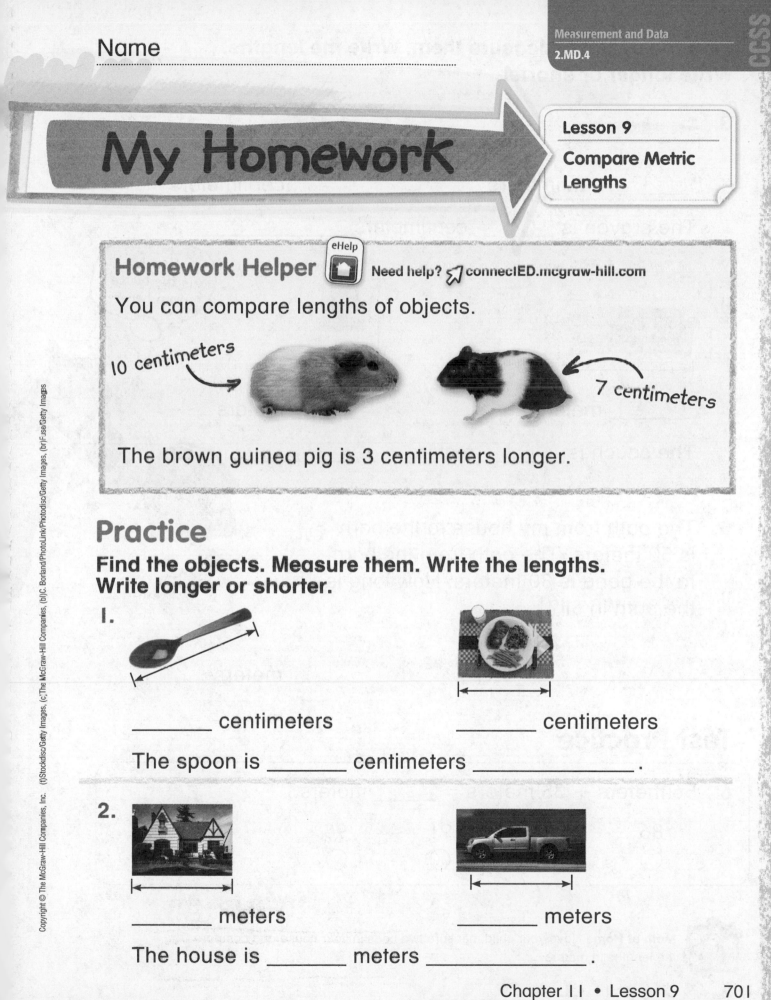

10 centimeters

7 centimeters

The brown guinea pig is 3 centimeters longer.

Practice

**Find the objects. Measure them. Write the lengths.
Write longer or shorter.**

1.

_____ centimeters   _____ centimeters

The spoon is _____ centimeters _____.

2.

_____ meters   _____ meters

The house is _____ meters _____.

**Find the objects. Measure them. Write the lengths.
Write longer or shorter.**

3.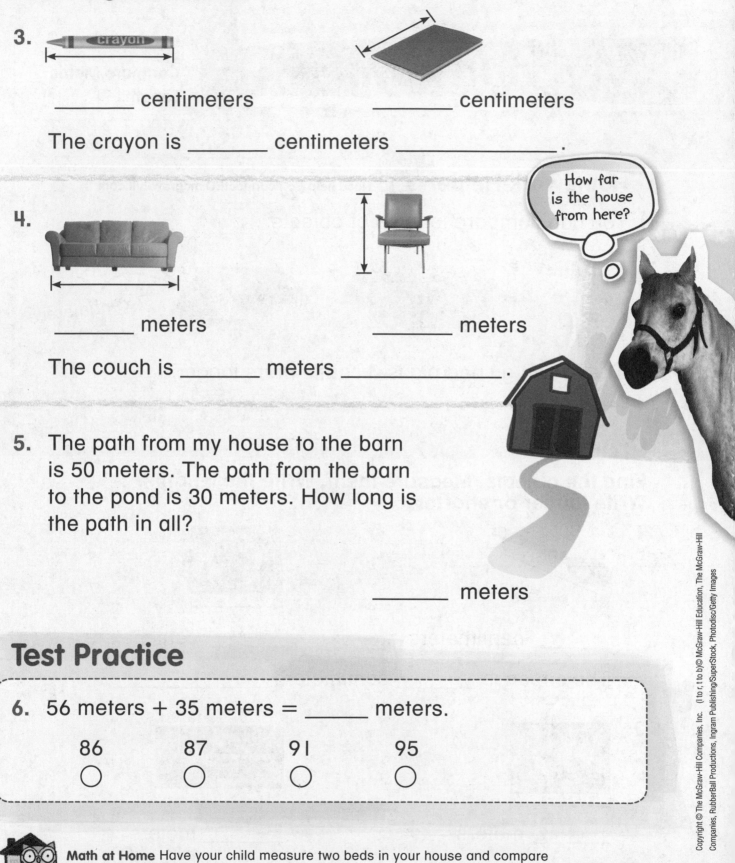

_____ centimeters   _____ centimeters

The crayon is _____ centimeters _____ .

4.

_____ meters   _____ meters

The couch is _____ meters _____ .

How far is the house from here?

5. The path from my house to the barn is 50 meters. The path from the barn to the pond is 30 meters. How long is the path in all?

_____ meters

Test Practice

6. 56 meters + 35 meters = _____ meters.

    86   87   91   95
    ○   ○   ○   ○

Math at Home Have your child measure two beds in your house and compare the lengths in meters.

Name

Relate Centimeters and Meters

Lesson 10

ESSENTIAL QUESTION
How can I measure objects?

Explore and Explain

_____ centimeters or _____ meters

Teacher Directions: Choose a large object from the picture such as a table or window. Find it in your school. Measure the object in centimeters. Measure the object in meters. Discuss the measurements.

Online Content at 🖱 **connectED.mcgraw-hill.com**

Chapter 11 • Lesson 10

See and Show

You can use different units of length to measure the same object.

The basketball hoop is about _____3_____ meters tall.

It is about ____300____ centimeters tall.

> **Helpful Hint**
> 1 meter = 100 centimeters

Want to see my "behind-the-back, nothing but net" shot?

Find the object. Measure the length of each object twice.

Object	Measure
1.	_____ centimeters _____ meters
2.	_____ centimeters _____ meters
3.	_____ centimeters _____ meters

Talk Math Which unit of measure gives you a more exact measurement?

Name _____

On My Own

Find the object. Measure the length of each object twice.

Object	Measure
4. 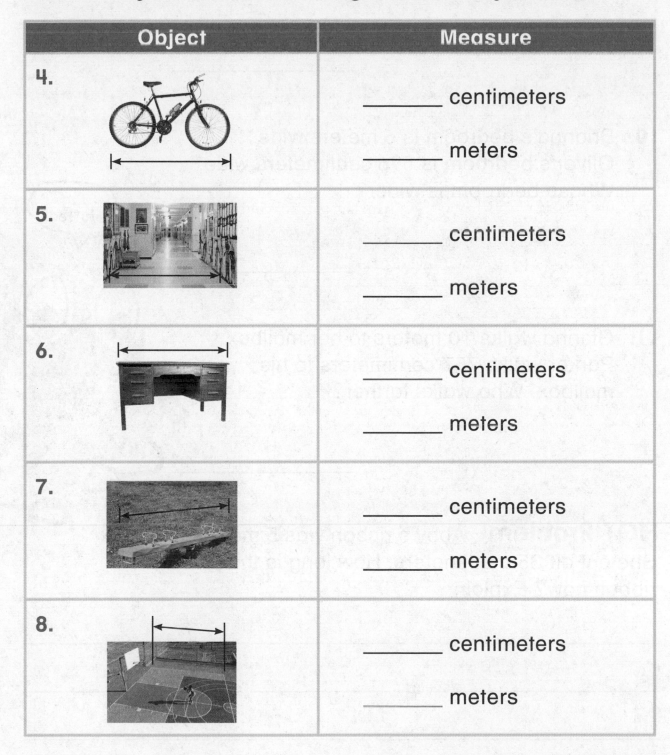	_____ centimeters _____ meters
5.	_____ centimeters _____ meters
6.	_____ centimeters _____ meters
7.	_____ centimeters _____ meters
8.	_____ centimeters _____ meters

Problem Solving

9. Kate's garden is 9 meters long.
Stella's garden is 954 centimeters
long. Whose garden is longer?

_____

10. Brianna's bedroom is 6 meters wide.
Oliver's bedroom is 596 centimeters wide.
Whose bedroom is wider?

_____

11. Gianna walks 10 meters to her mailbox.
Parker walks 757 centimeters to his
mailbox. Who walks farther?

_____

HOT Problem Abby's ribbon was 5 meters long.
She cut off 35 centimeters. How long is the piece of
ribbon now? Explain.

_____

_____

_____

_____

My Homework

Homework Helper

eHelp

Need help? connectED.mcgraw-hill.com

You can use different units of length
to measure the same object.

The yarn is 2 meters long.
It is 200 centimeters long.

Practice

Find the object. Measure the length of each object twice.

Object	Measure
1.	_____ centimeters _____ meters
2.	_____ centimeters _____ meters
3.	_____ centimeters _____ meters

Find the object. Measure the length of each object twice.

Object	Measure
4.	_____ centimeters _____ meters
5.	_____ centimeters _____ meters

6. At the end of the race, Carly was 3 meters ahead of Reagan. How many centimeters was she ahead of Reagan?

_____ centimeters

3 is my lucky number!

Test Practice

7. 35 centimeters + 65 centimeters = _____

    I centimeter   I meter   90 centimeters   2 meters

       ○   ○   ○   ○

 Math at Home Have your child measure an object in your house using centimeters and meters. Discuss how each measurement relates to the size of the unit.

Measure on a Number Line

Explore and Explain

Lesson 11

ESSENTIAL QUESTION
How can I measure objects?

I wonder where I'd be on a number line?

centimeters

| 0 | 1 | 2 | 3 | 4 | 5 | 6 | 7 | 8 | 9 | 10 | 11 | 12 | 13 | 14 | 15 |

_____ − _____ = _____

Teacher Directions: Place a crayon and a pair of scissors on the number line. Line up the ends at 0. How much longer are the scissors than the crayon? Repeat with other pairs of objects. Draw a pair of objects on the number line and write the subtraction sentence that shows how much longer one object is than another.

See and Show

centimeters

20 21 22 23 24 25 26 27 28 29 30

The crayon is _____**9**_____ centimeters long.

The paper clip is _____**5**_____ centimeters long.

Use the number line to answer the questions.

inches

60   61   62   63   64   65   66

1. How long is the pencil? _____ inches

2. How much longer is the marker than
   the rubber band? _____ inches

Talk Math How does the number line
help you compare measurements?

Name

On My Own

Use the number line to answer the questions.

3. How long is the glue stick? _____ centimeters

4. How long is the push pin? _____ centimeters

5. How long is the eraser? _____ centimeters

6. How much longer is the pencil
   than the glue stick? _____ centimeters

7. How long are the eraser and
   push pin in all? _____ centimeters

8. How much shorter is the push
   pin than the pencil? _____ centimeters

Problem Solving

Mathematical
PRACTICE

9. Makenna is 38 inches tall. Her brother is
26 inches tall. How much taller is Makenna?

See, I told you
I was longer!

_____ inches

10. Carly's dog Sebastian is 22 inches long.
Her cat, Annabelle, is 17 inches long.
How much longer is Sebastian?

_____ inches

Meow!

11. Levi found a trail in the woods that is
54 meters long. Then he found a second
trail that is 83 meters long. How much
longer is the second trail?

_____ meters

Write Math How can a number line help you measure
objects? Explain.

_____

_____

_____

Name

My Homework

Lesson 11

Measure on a Number Line

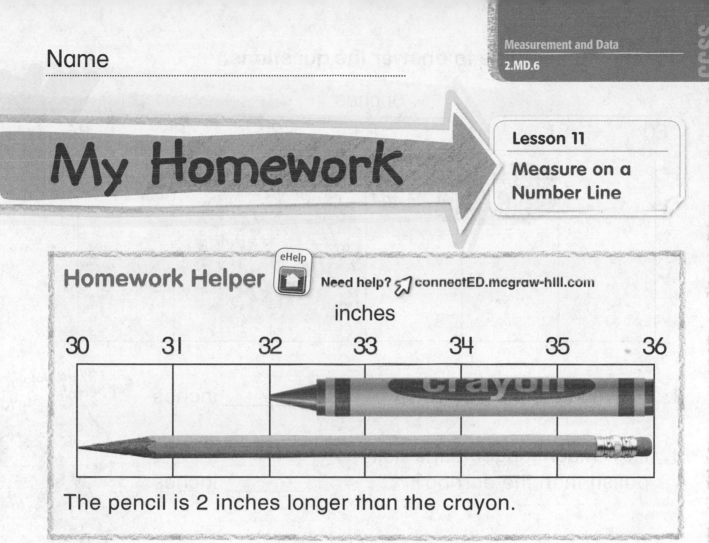

Homework Helper

Need help? connectED.mcgraw-hill.com

inches

| 30 | 31 | 32 | 33 | 34 | 35 | 36 |

The pencil is 2 inches longer than the crayon.

Practice

Use the number line to answer the questions.

centimeters

60 61 62 63 64 65 66 67 68 69 70 71 72 73 74 75

1. How long is the truck? _____ centimeters

2. How long is the car? _____ centimeters

Chapter 11 • Lesson 11   713

Use the number line to answer the questions.

inches

| 80 | 81 | 82 | 83 | 84 | 85 | 86 |

3. How long is the barrette? _____ inches

Favorite color:
Peppermint Stick!

4. How much longer is the nail
   polish than the earring? _____ inches

5. Sean builds a block tower that is 35 inches high.
   Victoria builds a block tower that is 27 inches
   high. How much higher is Sean's tower?

   _____ inches

Test Practice

6. 36 inches + 26 inches = _____

   ○ 62 inches   ○ 62 centimeters   ○ 62 feet   ○ 62 meters

Math at Home Help your child create a large number line measured
in inches. Have him or her measure objects from your home using the
number line.

Name

Measurement Data

Lesson 12

ESSENTIAL QUESTION
How can I measure objects?

Explore and Explain

I can measure anything!

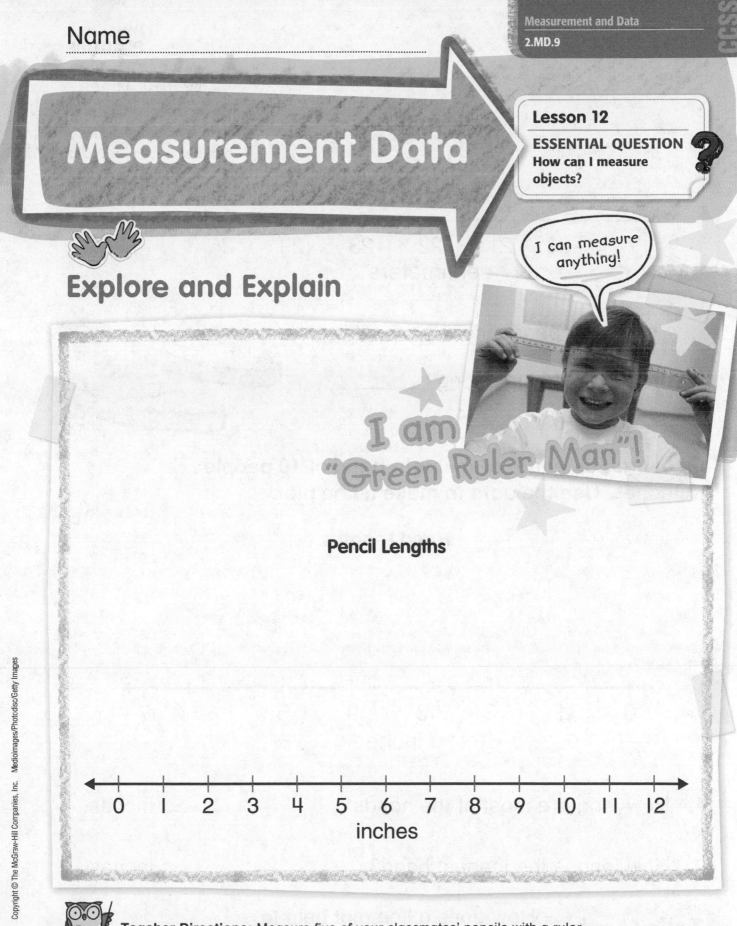

I am "Green Ruler Man"!

Pencil Lengths

0 1 2 3 4 5 6 7 8 9 10 11 12

inches

Teacher Directions: Measure five of your classmates' pencils with a ruler.
Make a line plot using the data from your measurements.

See and Show

Shoe Lengths

How many people have shoes that are 23 centimeters long? ____2____

Measure the length of the right hand of 10 people in inches. Use the data to make a line plot.

Hand Lengths

1. How long are most of the hands? _____ inches

2. How long is the longest hand? _____ inches

Talk Math How does a line plot help to show measurement data?

Name _____

On My Own

Measure 15 books from your classroom library in centimeters. Use the data to make a line plot.

Book Lengths

10 11 12 13 14 15 16 17 18 19 20 21 22 23 24 25
centimeters

3. How long are most books? _____ centimeters

This one's a whopper!

4. How long is the longest book? _____ centimeters

Measure 15 used crayons in inches. Use the data to make a line plot.

Crayon Lengths

0 1 2 3 4 5
inches

5. How short is the shortest crayon? _____ inches

6. How long is the longest crayon? _____ inches

Problem Solving

7. A line plot shows that 12 people are 48 inches tall, 8 people are 50 inches tall, and 9 people are 47 inches tall. How tall are most of the people?

_____ inches

Did they count my tongue?

8. Four snakes are 20 inches, 2 snakes are 32 inches, and 4 snakes are 15 inches. What is the difference in length between the longest and the shortest snake?

_____ inches

9. Four people went down the slide 8 times. Eight people went down the slide 6 times. Two people went down the slide 12 times. How many times did most people go down the slide?

_____ times

HOT Problem Jordan made a line plot of the heights of his 5 pets. The same number of pets are 14 inches and 18 inches tall. One pet is 6 inches tall. How many pets are 14 inches tall?

_____

_____

_____

Name ....

My Homework

Homework Helper

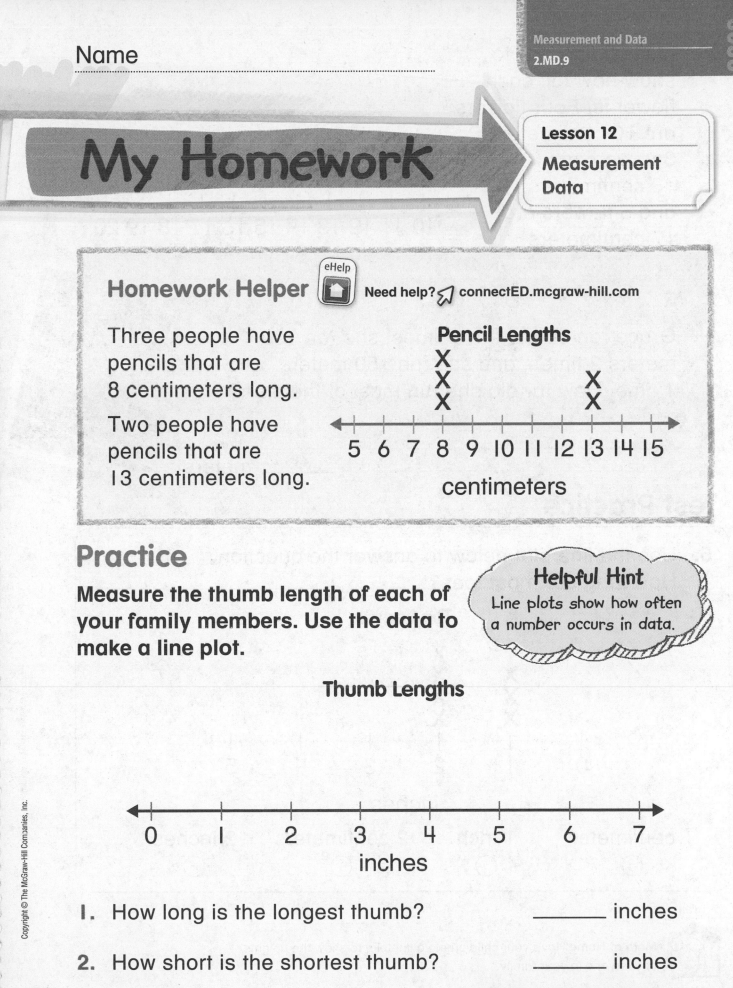

eHelp   Need help? connectED.mcgraw-hill.com

Three people have pencils that are 8 centimeters long.

Two people have pencils that are 13 centimeters long.

Pencil Lengths

X
X   X
X   X
5 6 7 8 9 10 11 12 13 14 15

centimeters

Practice

Measure the thumb length of each of your family members. Use the data to make a line plot.

Helpful Hint
Line plots show how often a number occurs in data.

Thumb Lengths

0   1   2   3   4   5   6   7

inches

1. How long is the longest thumb? _____ inches

2. How short is the shortest thumb? _____ inches

3. Make a line plot to
show how tall each
flower is. Four flowers
are 10 centimeters,
3 flowers are
13 centimeters,
and 6 flowers are
17 centimeters.

4. Grace ran 50 meters 3 times, she ran 100
meters 2 times, and she ran 150 meters
1 time. How far did she run most of the time?

_____ meters

Test Practice

5. Use the line plot below to answer the question.
How long are most toes?

1 centimeter   1 inch   2 centimeters   2 inches

○   ○   ○   ○

Math at Home Have your child create a line plot to show the heights
of everyone in your family.

My Review

Vocabulary Check

Write the correct word in each blank.

length	inch	estimate	measure
foot	yard	centimeter	meter

1. You can _____ to find a number close to an exact amount.

2. _____ is how long or how far away something is.

3. A paper clip is about one _____ long.

4. A _____ is equal to 3 feet.

5. To _____ is to find the length, height, or weight using standard or nonstandard units.

6. A base-ten cube is about one _____ long.

7. A _____ is equal to 12 inches.

8. A _____ is equal to 100 centimeters.

We only have 2 yards to go!

Concept Check

Find the object. Estimate the length. Measure the object in centimeters.

Object	Estimate	Measure
9.	about _____ centimeters	about _____ centimeters

Find the object. Choose the tool and measure it.

Object	Tool	Measure
10.	_____	about _____

Find the objects. Measure them. Write the lengths. Write longer or shorter.

11.

_____ centimeters   _____ centimeters

The glue stick is _____ centimeters _____.

Find the object. Measure the length of it twice.

Object	Measure
12.	_____ centimeters _____ meters

Name _____

Problem Solving

13. Use the data to make a line plot.
6 pieces of yarn were 4 inches long,
3 pieces of yarn were 2 inches long,
2 pieces of yarn were 6 inches long,
and 1 piece of yarn was 5 inches long.

Yarn Lengths

0   1   2   3   4   5   6
inches

14. Mason's bike is 4 feet long. His scooter is
2 feet long. How many inches longer is his
bike than his scooter?

_____ inches

Test Practice

15. Lucy's lacrosse stick is 1 meter long. How many
centimeters long is Lucy's lacrosse stick?

50   100   200   400
○   ○   ○   ○

Complete each sentence.

ESSENTIAL QUESTION How can I measure objects?

Customary	Metric
A foot is equal to _____ inches.	A meter is equal to _____ centimeters.
A yard is equal to _____ inches or _____ feet.	My foot is about _____ centimeters long.
My foot is about _____ inches long.	I can measure metric lengths with a _____ or a _____.
I can measure customary lengths with a _____, _____, or _____.	

We're good sports too!

We are winners!

Chapter 12

Geometric Shapes and Equal Shares

ESSENTIAL QUESTION

How do I use shapes and equal parts?

Let's Go to the Park!

Watch a video!

Watch

My Common Core State Standards

Geometry

2.G.1 Recognize and draw shapes having specified attributes, such as a given number of angles or a given number of equal faces. Identify triangles, quadrilaterals, pentagons, hexagons, and cubes.

2.G.2 Partition a rectangle into rows and columns of same-size squares and count to find the total number of them.

2.G.3 Partition circles and rectangles into two, three, or four equal shares, describe the shares using the words *halves*, *thirds*, *half of*, *a third of*, etc., and describe the whole as two halves, three thirds, four fourths. Recognize that equal shares of identical wholes need not have the same shape.

Standards for Mathematical PRACTICE

1. Make sense of problems and persevere in solving them.
2. Reason abstractly and quantitatively.
3. Construct viable arguments and critique the reasoning of others.
4. Model with mathematics.
5. Use appropriate tools strategically.
6. Attend to precision.
7. Look for and make use of structure.
8. Look for and express regularity in repeated reasoning.

= focused on in this chapter

Name

Draw an X over the object that is a different shape.

1.

2.

Draw a line to match the objects that are the same shape.

3.

4.

5.

6. What shape is the game board?
   Circle the word.

   triangle   square   circle

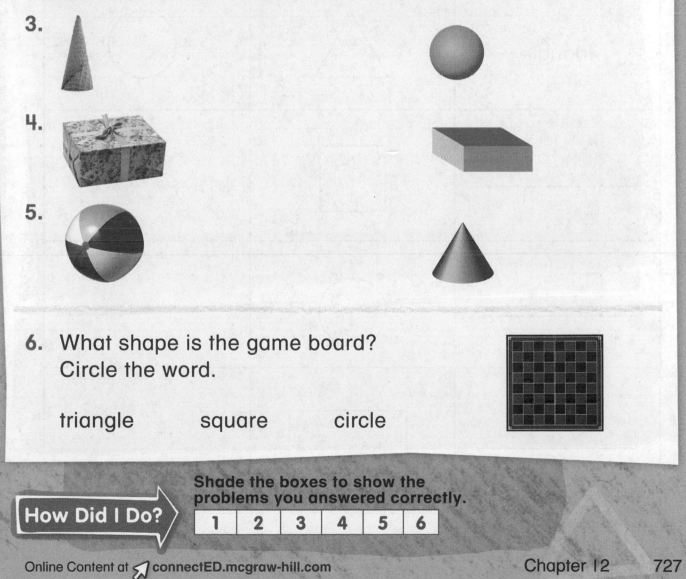

How Did I Do? ⟶

Shade the boxes to show the problems you answered correctly.

| 1 | 2 | 3 | 4 | 5 | 6 |

Online Content at connectED.mcgraw-hill.com   Chapter 12   727

Name ....

Review Vocabulary

circle	rectangle	square

Use the review vocabulary to complete the chart.
The first row is done for you.

Word	Example	Not an Example
triangle	△	○
_____	□	
_____	○	
_____	▭	

My Vocabulary Cards

Vocab

Lesson 12-2

angle

Lesson 12-4

cone

Lesson 12-4

cube

Lesson 12-4

cylinder

Crop Sweet Corn Whole Kernel

Lesson 12-5

edge

Lesson 12-5

face

Directions:
Ideas for Use

- Group 2 or 3 common words. Add a word that is unrelated to the group. Ask another student to name the unrelated word.

- Have students draw an example for each card. Have them make drawings that are different from what is shown on each card.

A three-dimensional shape that narrows to a point from a circular face.

Where two sides on a two-dimensional shape meet.

A three-dimensional shape that is shaped like a can.

A three-dimensional shape with 6 square faces.

The flat part of a three-dimensional shape.

The line segment where two faces of a three-dimensional shape meet.

My Vocabulary Cards

Mathematical
PRACTICE

Lesson 12-7

fourths

4 fourths

Lesson 12-7

halves

2 halves

Lesson 12-1

hexagon

Lesson 12-1

parallelogram

Lesson 12-7

partition

Lesson 12-1

pentagon

Directions:
More Ideas for Use
- Create a riddle for each word. Have a friend guess the word for each riddle.

- Have students write a tally mark on each card every time they read the word in this chapter or use it in their writing. Challenge them to try to use at least 10 tally marks for each word card.

Two equal parts.

Four equal parts.

A two-dimensional shape that has two pairs of sides that are the same length and are equal distance apart.

A two-dimensional shape that has six sides.

A polygon with five sides.

To divide into groups or "break up."

My Vocabulary Cards

Lesson 12-4

pyramid

Lesson 12-1

quadrilateral

Lesson 12-2

rectangular prism

Cereal

Lesson 12-4

side

Lesson 12-6

sphere

Lesson 12-1

thirds

3 thirds

Directions:
More Ideas for Use
- Have students sort the words by the number of syllables in each word.
- Ask students to find pictures to show each word. Have them work with a friend to guess which word the picture shows.

A shape that has 4 sides and 4 angles.

A three-dimensional shape with a polygon as a base and other faces that are triangles.

One of the line segments that make up a shape.

A three-dimensional shape with 6 faces that are rectangles.

Three equal parts.

A three-dimensional shape that has the shape of a round ball.

My Vocabulary Cards

Mathematical
PRACTICE

Lesson 12-4

three-dimensional shapes

Lesson 12-7

trapezoid

Lesson 12-1

two-dimensional shapes

Lesson 12-1

vertex

vertex

Directions:
More Ideas for Use
• Have students arrange the cards in alphabetical order.

• Group 2 or 3 common words. Add a word that is unrelated to the group. Ask another student to name the unrelated word.

A two-dimensional shape with four sides and only two opposite sides that are parallel.

A shape that has: length, width, and height.

The point where three or more faces of a three-dimensional shape meet.

A shape having the two dimensions length and width.

Name

Two-Dimensional Shapes

Lesson 1

ESSENTIAL QUESTION
How do I use shapes and equal parts?

Don't be such a square.

Explore and Explain

Watch ▶   Tools

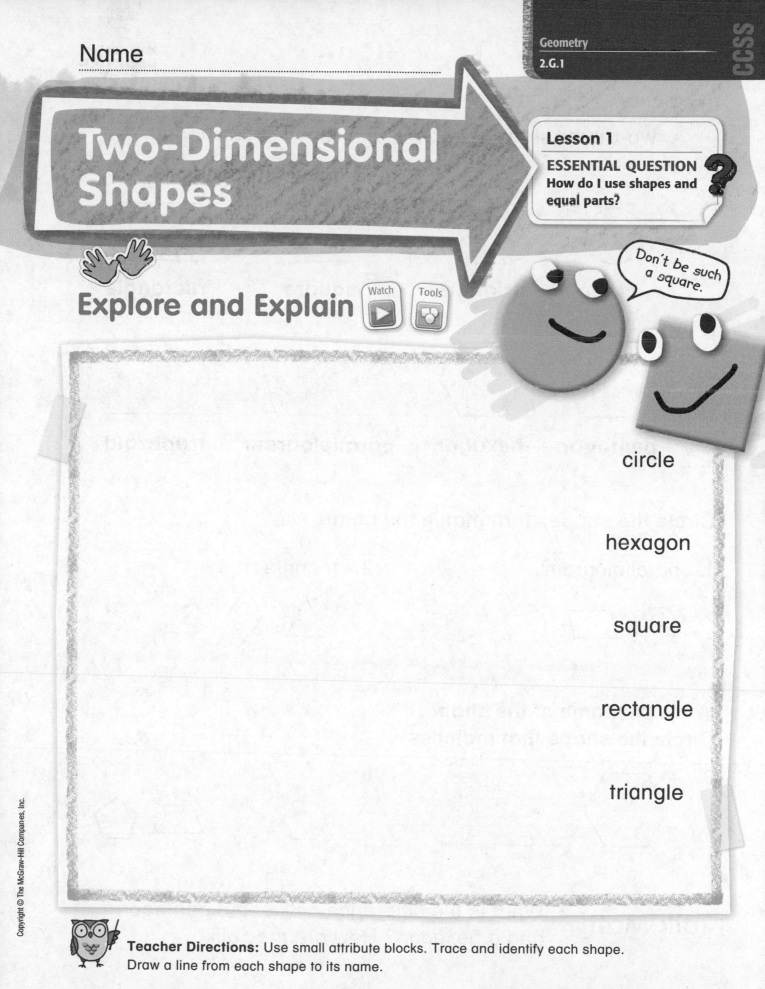

circle

hexagon

square

rectangle

triangle

Teacher Directions: Use small attribute blocks. Trace and identify each shape.
Draw a line from each shape to its name.

See and Show

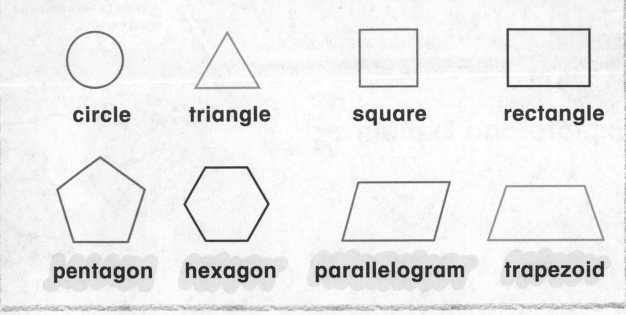

A **two-dimensional shape** is a shape with only length and width.

circle   triangle   square   rectangle

pentagon   hexagon   parallelogram   trapezoid

Circle the shapes that match the name.

1. parallelogram

2. triangle

Write the name of the shape.
Circle the shape that matches.

3. _____

Talk Math   What is the difference between a pentagon and a hexagon? How are they alike?

On My Own

Circle the shapes that match the name.

4. trapezoid

5. hexagon

6. triangle

7. pentagon

Write the name of the shape.
Circle the shape that matches.

8.

_____

9.

_____

Circle the shape that does not belong in each group.

10.

11.

Problem Solving

Mathematical PRACTICE

12. Identify the shape of each sign.

_____   _____

How many of each shape do you see?

13.

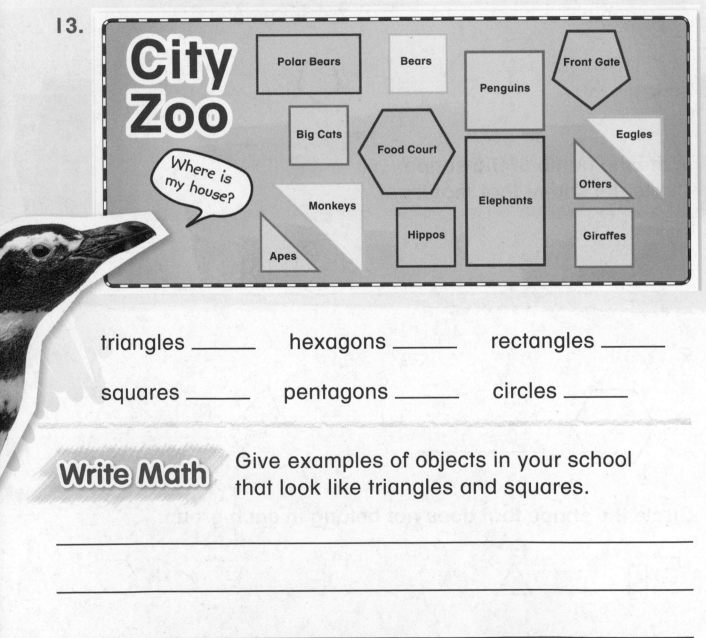

triangles _____   hexagons _____   rectangles _____

squares _____   pentagons _____   circles _____

Write Math   Give examples of objects in your school that look like triangles and squares.

_____

_____

_____

Name _____

My Homework

Lesson 1

Two-Dimensional Shapes

Homework Helper eHelp   Need help? connectED.mcgraw-hill.com

A two-dimensional shape is a shape with only length and width.

circle   triangle   square   rectangle

pentagon   hexagon   parallelogram   trapezoid

Practice

Circle the shapes that match the name.

1. rectangle

2. triangle

3. trapezoid

4. hexagon

Write the name of the shape. Circle the shape that matches.

5. _____

6. _____

7. Jack cut out a shape to glue onto a picture. The shape looked like an ice cream cone. What shape did he cut out?

   _____

mmm, strawberry!

Vocabulary Check

8. **Circle the hexagons.**

Math at Home Point to two-dimensional shapes around your house (triangles, squares, rectangles, hexagons, and pentagons) and have your child identify each shape.

Name _____

Sides and Angles

Explore and Explain

Watch Tools

Did you see what I saw?

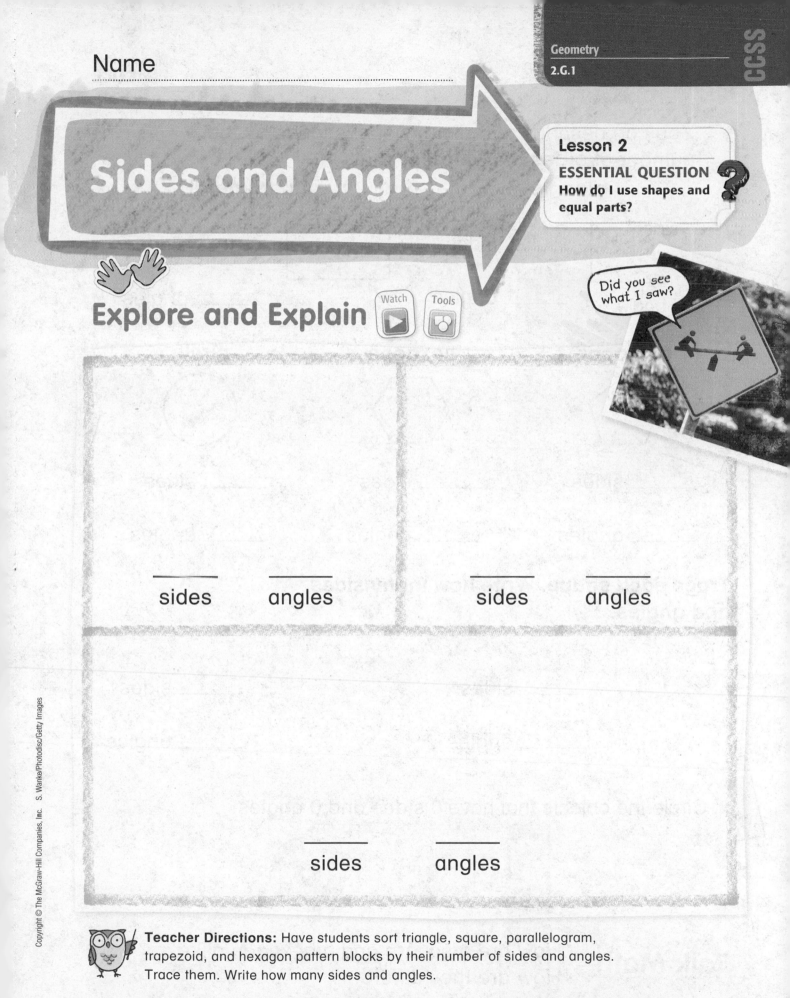

_____ _____   _____ _____
sides angles   sides angles

_____ _____
sides angles

Teacher Directions: Have students sort triangle, square, parallelogram, trapezoid, and hexagon pattern blocks by their number of sides and angles. Trace them. Write how many sides and angles.

See and Show

You can describe two-dimensional shapes by the number of **sides** and **angles**.

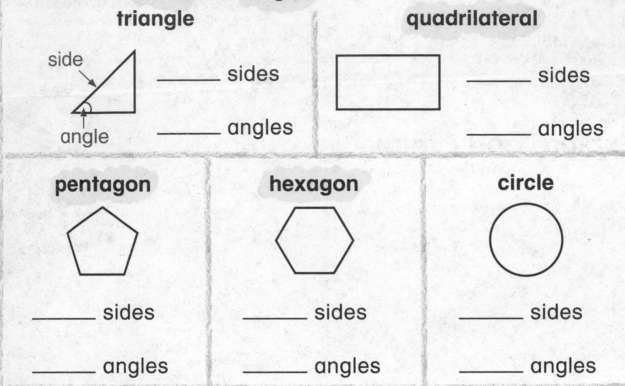

triangle

side

angle

_____ sides

_____ angles

quadrilateral

_____ sides

_____ angles

pentagon

_____ sides

_____ angles

hexagon

_____ sides

_____ angles

circle

_____ sides

_____ angles

Trace each shape. Write how many sides and angles.

1. _____ sides

_____ angles

2. _____ sides

_____ angles

3. Circle the objects that have 0 sides and 0 angles.

  How are a square and a hexagon alike?
How are they different?

Name _____

On My Own

Trace each shape. Write how many sides and angles.

4. _____ sides

_____ angles

5. _____ sides

_____ angles

6. _____ sides

_____ angles

7. _____ sides

_____ angles

Circle the objects that match the description.

8. 3 sides and 3 angles

9. 4 sides and 4 angles

Problem Solving

Draw a picture to solve.

10. Kira draws a shape with 6 sides and
6 angles. What shape does she draw?

6 sides!

_____

11. Alex draws a shape with 3 sides and 3 angles.
What shape does Alex draw?

_____

12. Josh drew 3 squares. Katie drew 2 triangles
and 1 square. Who drew more angles?

_____

Write Math   Write the name of each shape.
Describe two things about each shape.

1. _____

2. _____

1. _____

2. _____

Name

My Homework

Lesson 2
Sides and Angles

Homework Helper

eHelp Need help? connectED.mcgraw-hill.com

A two-dimensional shape can be described by its sides and angles.

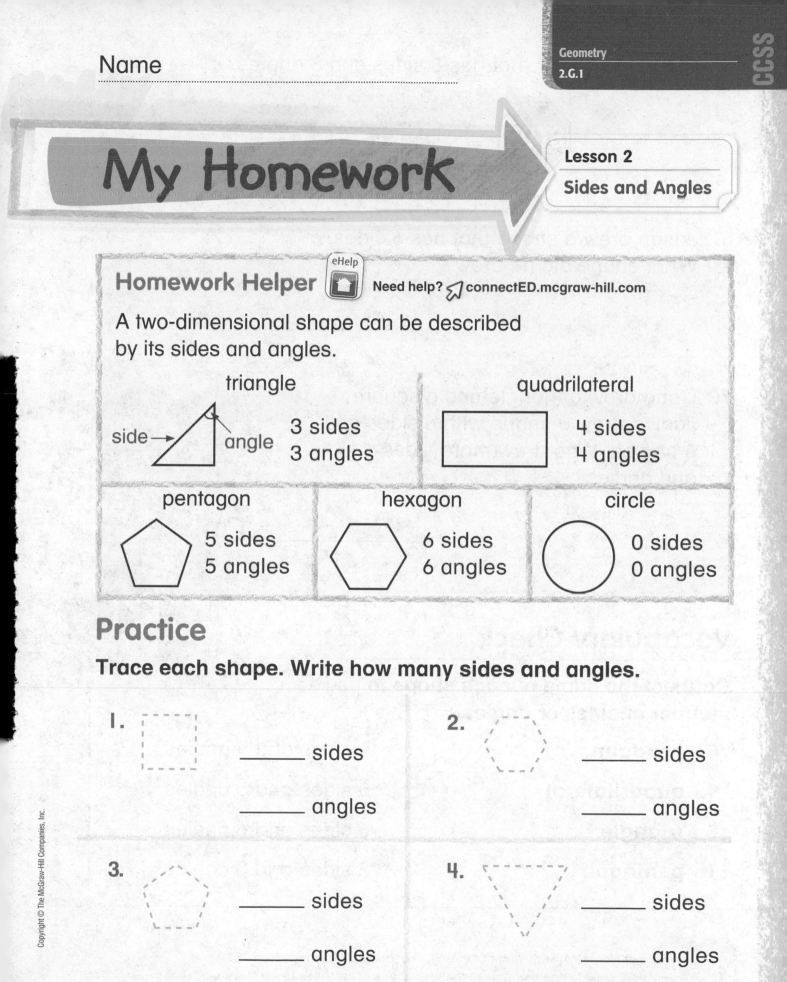

triangle

side → angle

3 sides
3 angles

quadrilateral

4 sides
4 angles

pentagon

5 sides
5 angles

hexagon

6 sides
6 angles

circle

0 sides
0 angles

Practice

Trace each shape. Write how many sides and angles.

1. _____ sides

_____ angles

2. _____ sides

_____ angles

3. _____ sides

_____ angles

4. _____ sides

_____ angles

5. Circle the object that has 8 sides and 8 angles.

6. Jason drew a shape that has 6 sides. What shape did he draw?

_____

7. Carla drew a triangle and a square. Janice drew a shape with 6 sides and 6 angles. Who drew more sides and angles?

_____

Circle time!

Vocabulary Check Vocab abc

Connect the name of each shape to its number of sides or angles.

8. hexagon	4 sides and 4 angles
9. quadrilateral	5 sides and 5 angles
10. triangle	6 sides and 6 angles
11. pentagon	3 sides and 3 angles

Math at Home While driving or walking, look at road signs together. Ask your child to name and describe the shapes of the signs he or she sees.

Copyright © The McGraw-Hill Companies, Inc.   (tl)Ryan McVay/Photodisc/Getty Images, (tc)The McGraw-Hill Companies, (b)Digital Vision/Getty Images

Name ....

Problem Solving

STRATEGY: Draw a Diagram

Lesson 3

ESSENTIAL QUESTION
How do I use shapes
and equal parts?

Lyla drew a shape. The shape
has 6 sides. It also has 6 angles.
What shape did Lyla draw?

Watch Tools

This way to the
park soccer field.

I have 6 sides!

I have 5!

1 Understand Underline what you know
Circle what you need to find.

2 Plan How will I solve the problem?

3 Solve Draw a diagram.

Lyla drew a _____.

4 Check Is my answer reasonable? Explain.

Practice the Strategy

Marcy drew a shape.
It has 5 sides. It has 5 angles.
What shape did she draw?

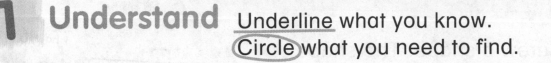

1 Understand Underline what you know.
Circle what you need to find.

2 Plan How will I solve the problem?

3 Solve I will...

Marcy drew a _____.

4 Check Is my answer reasonable? Explain

Drawing at the park is fun!

Name _____

Apply the Strategy

1. If a shape has 3 sides and 3 angles, what shape is it? Draw the shape.

   _____

2. Abby draws a triangle. Samuel draws a shape that has 1 more side than a triangle. What shape did Samuel draw? Draw the shape.

   _____

3. Jason drew a shape that has more sides than a triangle or rectangle but less angles than a hexagon. What shape did he draw? Draw the shape.

   _____

Choose a strategy
- Write a number sentence.
- Draw a diagram.
- Use logical reasoning.

4. Tammy saw a two-dimensional shape. The shape has 6 sides and 6 angles. Two of the sides are longer than the others. What shape did Tammy see?

_____

5. The sign at the end of David's street is a two-dimensional shape. It has 4 sides and 4 angles. The sides are all the same length. What shape is the sign?

_____

6. Jason was drawing shapes with sidewalk chalk in the park. He drew 3 triangles and then 2 squares. How many angles did he draw?

_____ angles

My Homework

Jacob was looking for shapes in the stars. He found one with 4 equal length sides and 4 angles.
What shape did Jacob see in the stars?

eHelp

1 Understand Underline what you know.
Circle what you need to find.

2 Plan How will I solve the problem?

3 Solve Draw a diagram.

Jacob saw a square.

4 Check Is my answer reasonable? Explain.

Underline what you know. Circle what you need to find. Draw a diagram to solve.

Home, sweet home!

1. Maggie drew a house. She drew a square for the bottom. She drew a triangle on top of the square for the roof. Trace the outside of the house. What shape is Maggie's house? Draw it.

_____

2. Billy saw a sign while walking through the park. The sign had no sides and no angles. What shape is the sign?

_____

3. Landon painted a shape that had 4 sides. The shape had 4 angles. What kind of shape did he paint?

_____

 Math at Home Describe a shape to your child. Have him or her draw the shape you described and identify the shape.

Check My Progress

Vocabulary Check [Vocab]

Complete each sentence.

angle	hexagon	pentagon
side	triangle	two-dimensional shape

1. A _____ has 5 sides and 5 angles.

2. A _____ has 3 sides and 3 angles.

3. A _____ has 6 sides and 6 angles.

Concept Check [Check]

Circle the shape or shapes that match the name.

4. triangle

5. pentagon

6. hexagon

7. quadrilateral

Write the name of the shape. Circle the shapes that match.

8. _____

9. _____

Write how many sides and angles.

10. _____ sides
_____ angles

11. _____ sides
_____ angles

12. _____ sides
_____ angles

13. _____ sides
_____ angles

Test Practice

14. Look at the shapes. Mark the shape that does not belong.

Three-Dimensional Shapes

Lesson 4

ESSENTIAL QUESTION
How do I use shapes
and equal parts?

Explore and Explain

Watch ▶ Tools

Teacher Directions: Look at the picture. Circle the three-dimensional shapes
that you see. Identify and describe each shape.

See and Show

A **three-dimensional shape** is a shape with length, width, and height.

sphere   cube   pyramid

cone   cylinder   rectangular prism

Write the name of the shape. Circle the objects that are the same shape.

1. _____

2. _____

Talk Math Name two objects in your classroom that are the same shape as a rectangular prism.

Name _____

My favorite sphere!

On My Own

**Write the name of the shape.
Circle the objects that are the
same shape.**

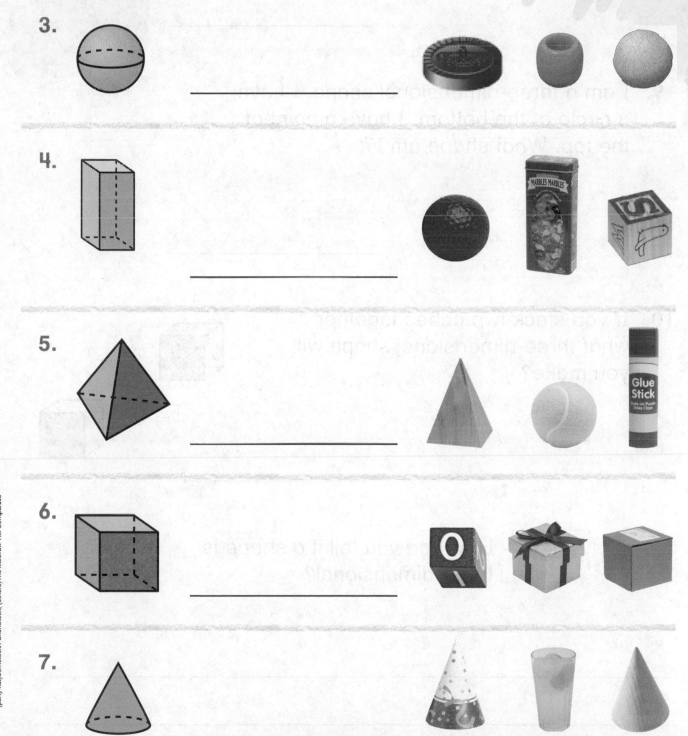

3. _____

4. _____

5. _____

6. _____

7. _____

Problem Solving

Mathematical
PRACTICE

8. Jen is wrapping a present that is in a box. The box is square on all sides. What shape is the box Jen is wrapping?

_____

9. I am a three-dimensional shape. I have a circle at the bottom. I have a point at the top. What shape am I?

_____

10. If you stack two cubes together what three-dimensional shape will you make?

_____

 Write Math   How can you tell if a shape is three-dimensional?

_____

_____

_____

My Homework

Homework Helper

eHelp

Need help? connectED.mcgraw-hill.com

A three-dimensional shape has length, width, and height.

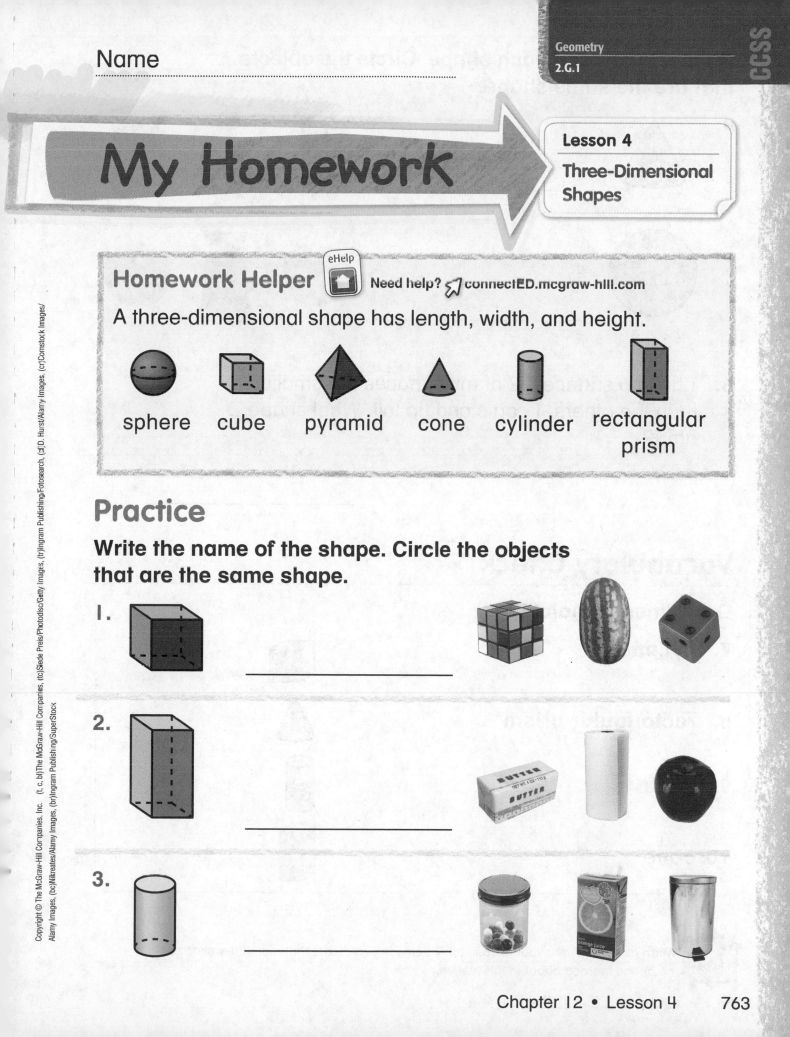

sphere   cube   pyramid   cone   cylinder   rectangular prism

Practice

Write the name of the shape. Circle the objects that are the same shape.

1. _____

2. _____

3. _____

Write the name of each shape. Circle the objects that are the same shape.

4.

   _____

5. _____

6. I have 6 surfaces. 2 of my surfaces are smaller than the others. I can stand up tall. What shape am I?

   _____

Vocabulary Check

Draw lines to match.

7. **cylinder**

8. **rectangular prism**

9. **cube**

10. **cone**

Math at Home Have your child identify items in your home that match the shapes he or she learned about in this lesson.

Faces, Edges, and Vertices

Lesson 5

ESSENTIAL QUESTION
How do I use shapes and equal parts?

Explore and Explain

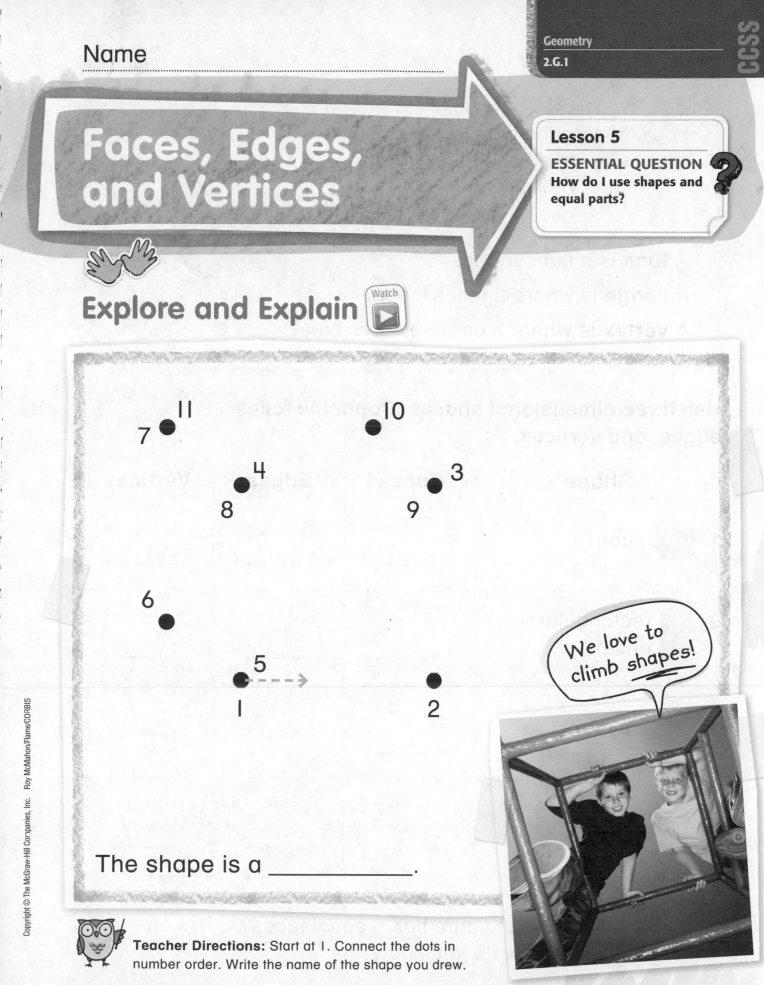

The shape is a _____.

Teacher Directions: Start at 1. Connect the dots in number order. Write the name of the shape you drew.

See and Show

You can describe three-dimensional shapes by the number of faces, edges, and vertices.

edge →

face

A **face** is a flat surface.

An **edge** is where 2 faces meet.

A **vertex** is where 3 or more faces meet.

vertex

Use three-dimensional shapes. Count the faces, edges, and vertices.

Shape	Faces	Edges	Vertices
1. cube	_____	_____	_____
2. rectangular prism	_____	_____	_____
3. pyramid	_____	_____	_____
4. sphere	_____	_____	_____

Talk Math What figure has 6 equal faces? How do you know?

Name _____

On My Own

Circle the shapes that match the description.

5. 0 faces, 0 edges, 0 vertices

6. 6 faces, 12 edges, 8 vertices

7. 5 faces, 8 edges, 5 vertices

8. 6 equal faces, 12 edges, 8 vertices

Circle the objects that match the description.

9. 6 faces, 12 edges, 8 vertices

10. 0 faces, 0 edges, 0 vertices

11. 6 faces, 12 edges, 8 vertices

12. 5 faces, 8 edges. 5 vertices

Problem Solving

13. Mindy drew the three shapes below. Circle the shape that has 6 faces and 12 edges.

Take a sphere to the park! We're lots of fun!

14. Ryan has a poster. The poster shows all of the shapes that have less than three faces. Circle the shape that is not on the poster.

HOT Problem Which shape does not belong? Circle it. Explain why it does not belong.

_____

_____

_____

My Homework

Lesson 5

Faces, Edges, and Vertices

Homework Helper

Need help? connectED.mcgraw-hill.com

Three-dimensional shapes are described
by the number of faces, edges, and vertices.

A face is a flat surface.

An edge is where 2 faces meet.

A vertex is where 3 or more faces meet.

edge → face

vertex

Practice

Circle the shapes or objects that matches the description.

1. 6 faces, 12 edges,
   8 vertices

2. 0 faces, 0 edges, 0 vertices

3. 5 faces, 8 edges,
   5 vertices

4. 6 faces, 12 edges,
   8 vertices

Circle the objects that match the descriptions.

5. 6 faces, 12 edges, 8 vertices

6. 0 faces, 0 edges, 0 vertices

7. I am a three-dimensional shape. I have 5 faces. I have 8 edges and 5 vertices. What shape am I?

> I love parties in the park!

Vocabulary Check

Complete each sentence.

face   edge   vertex

8. A _____ is a flat surface.

9. A _____ is where 3 or more faces meet.

10. An _____ is where 2 faces meet.

Math at Home Have your child identify real-life objects in your home that have the same shape as one of the shapes learned in this lesson.

770   Chapter 12 • Lesson 5

Relate Shapes and Solids

Lesson 6

ESSENTIAL QUESTION
How do I use shapes and equal parts?

Explore and Explain

Watch ▶ Tools

Yummy cubes!

Teacher Directions: Trace one face of a cube. Identify the shape. Trace the other faces of the cube. Describe the faces of a cube.

See and Show

The faces of three-dimensional shapes are two-dimensional shapes.

Helpful Hint
A cube has 6 equal faces. The faces are squares.

Circle the shapes that make the three-dimensional shape.

1.

2.

3.

Talk Math Explain how two-dimensional shapes and three-dimensional shapes are related.

On My Own

Circle the shapes that make the three-dimensional shape.

4.

5.

Circle the shape made by the faces.

6.

7.

8. Which of these shapes does not have a square as one of its faces?

Problem Solving

9. I have 6 equal faces. I have 8 vertices. What shape am I?

_____

I have only one face!

10. I have no faces and no vertices. What shape am I?

_____

11. Allison wants to trace a circle. Which objects could she use? Circle the objects.

Write Math Describe the faces that make a pyramid.

_____

_____

_____

Name _____

My Homework

Lesson 6

Relate Shapes and Solids

Homework Helper

eHelp

Need help? connectED.mcgraw-hill.com

The faces of three-dimensional shapes are two-dimensional shapes.

Helpful Hint
A rectangular prism has 4 rectangles and 2 squares as faces.

Practice

Circle the shapes that make the three-dimensional shape.

1.

2.

3.

Circle the shape made by the faces.

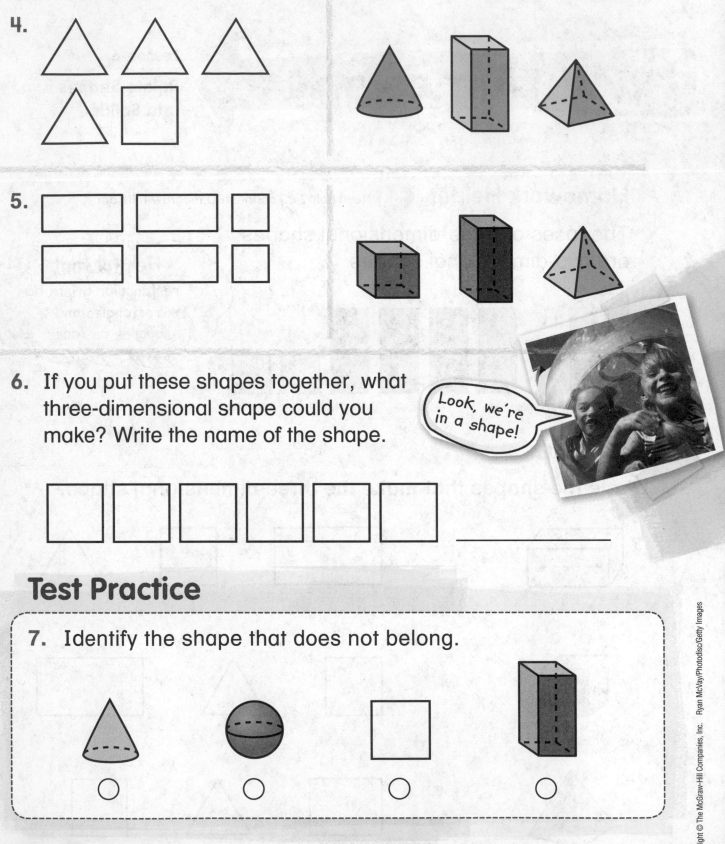

4.

5.

6. If you put these shapes together, what three-dimensional shape could you make? Write the name of the shape.

Look, we're in a shape!

Test Practice

7. Identify the shape that does not belong.

○   ○   ○   ○

Math at Home Ask your child to find an object at home that he or she could use to trace a rectangle on a piece of paper. Challenge your child to see if he or she can find something to trace for a circle.

Name _____

Halves, Thirds, and Fourths

Lesson 7

ESSENTIAL QUESTION
How do I use shapes and equal parts?

Explore and Explain 🛠 Tools

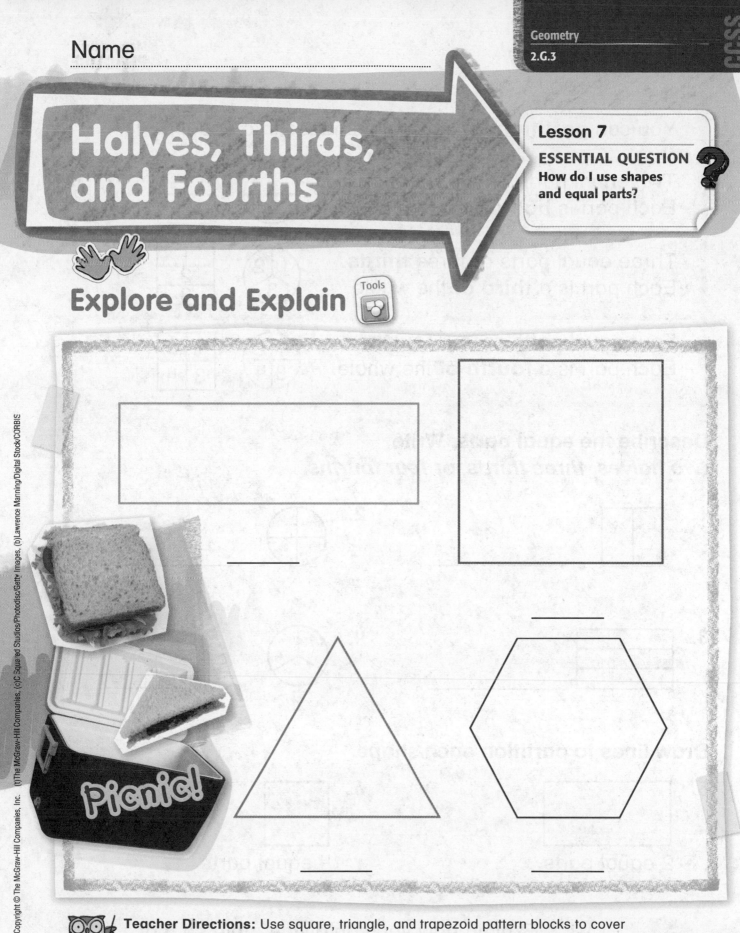

🦉 **Teacher Directions:** Use square, triangle, and trapezoid pattern blocks to cover each shape. Trace the blocks to show the shapes you used. Write how many blocks you used to cover each shape.

See and Show

You can **partition**, or separate, shapes into equal parts.

Two equal parts or two **halves.**
Each part is **half of** the whole.

Three equal parts or three **thirds.**
Each part is a **third of** the whole.

Four equal parts or four **fourths.**
Each part is a **fourth of** the whole.

Describe the equal parts. Write
two halves, *three thirds*, or *four fourths*.

1. _____

2. _____

3. _____

4. _____

Draw lines to partition each shape.

5.

2 equal parts

6.

4 equal parts

Talk Math   Explain how you can divide a pie so that four people each get an equal part.

Name

On My Own

Describe the equal parts. Write
two halves, three thirds, or four fourths.

7. _____

8. _____

Draw lines to partition each shape.

9.

4 equal parts

10.

2 equal parts

11.

3 equal parts

12.

2 equal parts

Partition the shape in a different way.
Show the same number of equal parts.

13.

14.

15.

16.

17. Eva's mom bought a pizza. Eva ate one equal part. Her friend ate one equal part. There was one equal part left for Eva's mom. How much of the pizza was left for Eva's mom?

_____ of the pizza

What's for lunch?

18. Gracie had a round slice of watermelon. She and her sister shared the slice equally. How much did each girl eat?

_____ of the watermelon slice

19. Sadie is making a picture for her cousin. She folds a piece of paper in half. Then she folds it in half again. She opens the paper. How many equal parts are there?

_____

HOT Problem Show the same number of equal parts in two different ways.

_____

_____

My Homework

Lesson 7

Halves, Thirds, and Fourths

Homework Helper eHelp Need help? connectED.mcgraw-hill.com

You can partition, or separate, shapes into equal parts.

halves | thirds | fourths

Practice

Describe the equal parts. Write *two halves, three thirds, or four fourths.*

1. _____

2. _____

3. _____

4. _____

Draw lines to partition each shape.

5. ⬜
3 equal parts

6. ⭕
2 equal parts

7. ▭
4 equal parts

Sharing halfsies is fun!

8. Nora and Brooke are sharing a sandwich. They each have an equal part. How much of the sandwich does each girl have?

_____ sandwich

Partition the shape in a different way.
Show the same number of equal shares.

9. ▦ ▭

10. ⊞ ⬜

Vocabulary Check

Color each shape as described.

11. ⭕
one half green

12. ⬜
one fourth blue

13. ▭
one third red

Copyright © The McGraw-Hill Companies, Inc.   Pixland/Jupiter Images

🏠 **Math at Home** Cut your child's food into either halves, thirds, or fourths. Ask him or her to identify how many equal parts you have created.

Name

Area

Lesson 8

ESSENTIAL QUESTION
How do I use shapes
and equal parts?

Explore and Explain

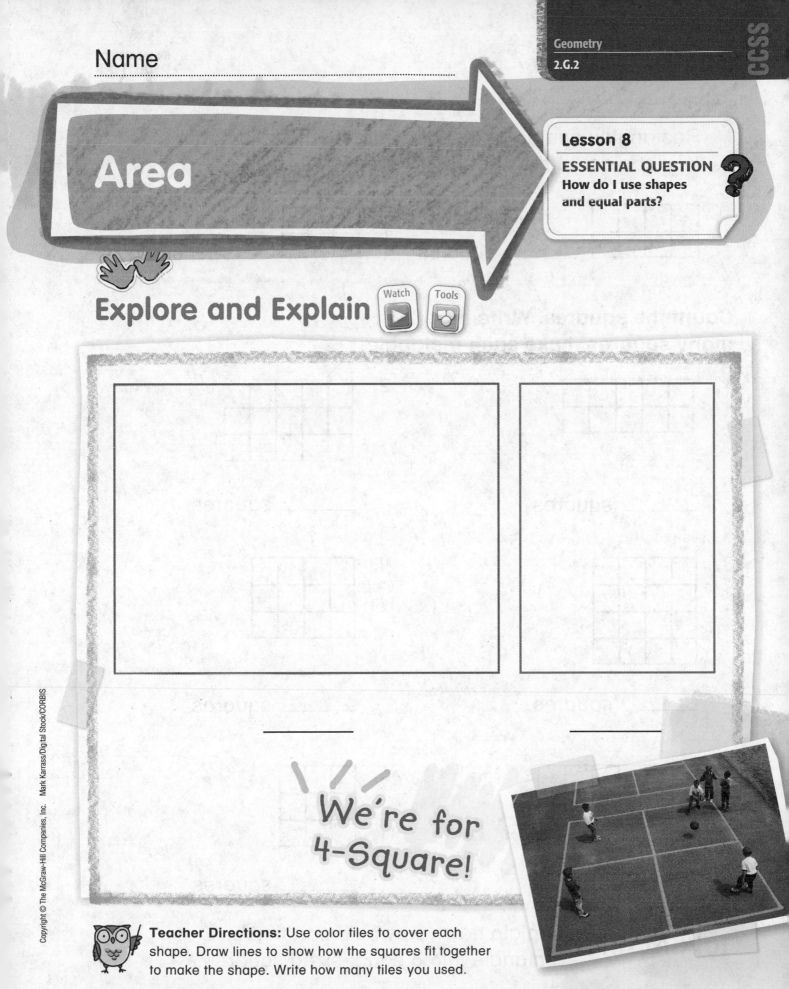

We're for
4-Square!

Teacher Directions: Use color tiles to cover each shape. Draw lines to show how the squares fit together to make the shape. Write how many tiles you used.

See and Show

Rectangles can be partitioned into equal-sized squares. You can count the squares to describe its size.

___8___ squares

___12___ squares

Count the squares. Write how many squares make each rectangle.

1. _____ squares

2. _____ squares

3. _____ squares

4. _____ squares

5. _____ squares

6. _____ squares

Talk Math Explain how you would partition a rectangle into 6 equal-sized squares.

Name _____

On My Own

Count the squares.
Write how many squares make each rectangle.

7.

_____ squares

8.

_____ squares

Use color tiles to cover each rectangle.
Write how many tiles were used.

9.

_____ tiles

10.

_____ tiles

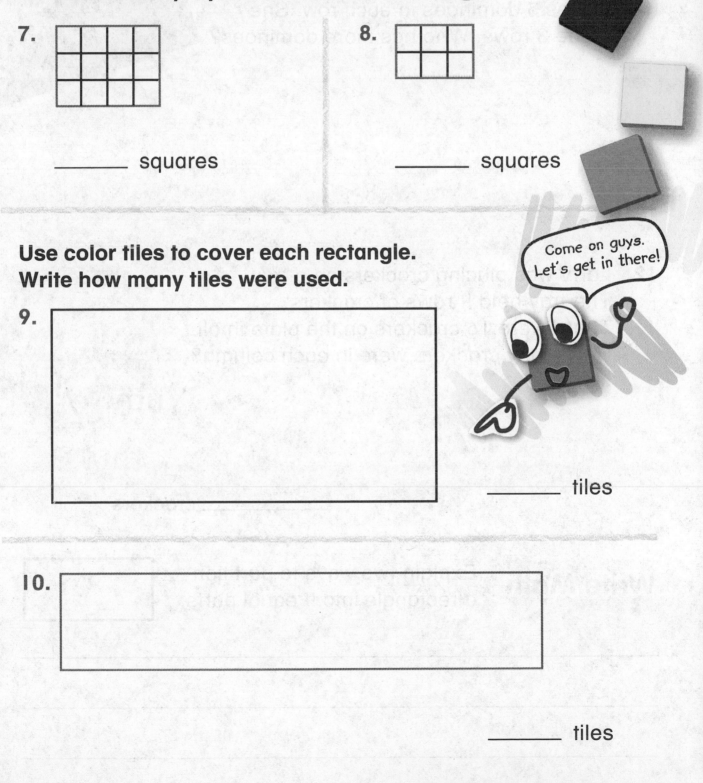

Come on guys. Let's get in there!

Problem Solving

Draw a picture to solve.

11. Addison and Corin each have a group of dominoes. Addison places 4 dominoes in each row. She made 5 rows. Corin places 5 dominoes in each row. She made 3 rows. Who has more dominoes?

_____

12. Larry was placing crackers on a tray.
The tray held 4 rows of crackers.
There were 16 crackers on the plate in all.
How many crackers were in each column?

Yummy!

_____ crackers

Write Math   Explain two ways to partition a rectangle into 4 equal parts.

_____

_____

Name _____

My Homework

Lesson 8

Area

Homework Helper

Need help? connectED.mcgraw-hill.com

A rectangle can be partitioned into squares to describe its size.

This rectangle is partitioned into 10 squares.

This rectangle is partitioned into 12 squares.

Practice

Count the squares. Write how many squares make each rectangle.

1. _____ squares

2. _____ squares

3. _____ squares

4. _____ squares

Count the squares. Write how many squares make each rectangle.

Form a straight line and <u>no</u> pushing!

5.

_____ squares

6.

_____ squares

7.

_____ squares

Draw a picture to solve.

8. Jan is cutting a rectangular pan of brownies. She cut them in half and then cut each of those halves in half again. She did the same thing going the other direction. How many brownies does she have?

_____ brownies

Test Practice

9. Choose the rectangle below that is partitioned using the greatest number of squares.

○   ○   ○   ○

Math at Home Find an opportunity to have your child help you determine how to cut something you are going to serve such as a casserole, brownies, cake, or rice treats. Together determine how many equal pieces you want and how to cut it.

Name _____

My Review

Vocabulary Check

Draw each shape.

1.   hexagon

2.   quadrilateral

Match each word to the correct shape.

3. cone

4. pyramid

5. sphere

6. cube

7. cylinder

8. rectangular prism

Concept Check

Circle the shapes that match the name.

9. rectangle

10. pentagon

Circle the shape that matches the description.

11. 3 sides and 3 angles

12. 6 sides and 6 angles

Write the name of the shape. Color the shapes that match.

13. _____

Look, it's my twin!

Circle the object that matches the description.

14. 6 faces, 12 edges, 8 vertices

Problem Solving

15. Caden's family is having pizza for dinner. There are 4 people in Caden's family. Draw lines to show how Caden's family should cut the pizza so that everyone gets an equal part.

16. Jamie drew a rectangle. He wants to partition it into equal-sized squares. He starts to partition the rectangle. Finish partitioning his rectangle. Write the total number of squares.

_____ squares

Test Practice

17. Blake has a peanut butter and jelly sandwich for lunch. He wants to share it with his 2 friends. Circle the word that tells how Blake should partition his sandwich.

halves   thirds   fourths   fifths

○   ○   ○   ○

Show the ways you can use shapes and equal parts.

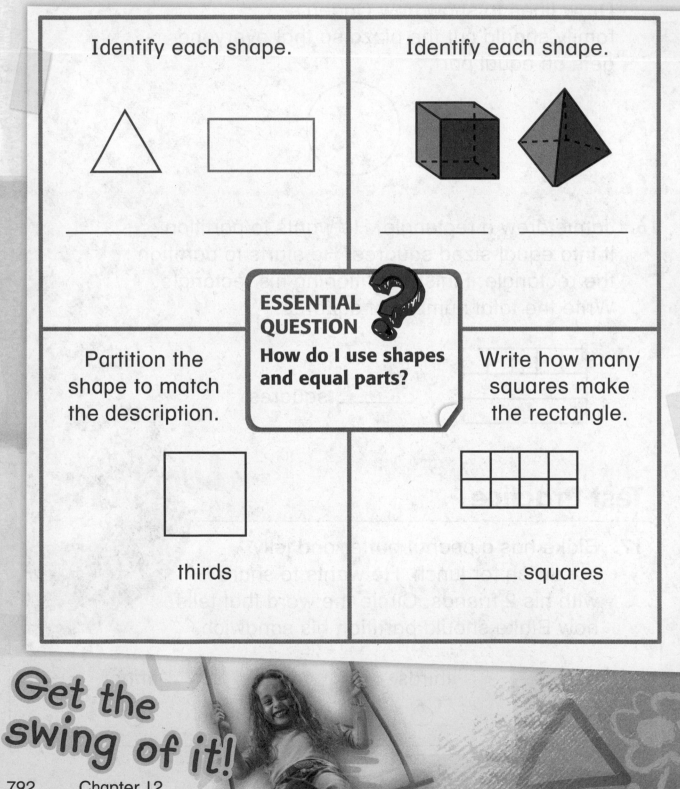

Identify each shape.

_____

Identify each shape.

_____ _____

ESSENTIAL QUESTION

How do I use shapes and equal parts?

Partition the shape to match the description.

thirds

Write how many squares make the rectangle.

_____ squares

Get the swing of it!

Glossary/Glosario

Vocab ← Go online for the eGlossary.

	English	**Spanish/Español**

A.M. The hours from midnight until noon.

a.m. Las horas que van desde la medianoche hasta el mediodía.

---

add (addition) Join together sets to find the total or sum. The opposite of *subtract*.

$$2 + 5 = 7$$

sumar (adición) Unir conjuntos para hallar el total o la suma. Lo opuesto de restar.

$$2 + 5 = 7$$

---

addend Any numbers or quantities being added together.

In 2 + 3 = 5, 2 is an addend and 3 is an addend.

$$2 + 3 = 5$$
↑ ↑

sumando Numeros o cantidades que se suman.

En 2 + 3 = 5, 2 es un sumando y 3 es un sumando.

$$2 + 3 = 5$$
↑ ↑

Aa

after Follow in place or time.

6 is just *after* 5

después Que sigue en lugar o en tiempo.

6 está justo *después* del 5

analog clock A clock that has an hour hand and a minute hand.

minute hand — hour hand

reloj analógico Reloj que tiene una manecilla horaria y un minutero.

minutero — manecilla horaria

angle Two sides on a two-dimensional shape meet to form an angle.

ángulo Dos lados de una figura bidimensional se encuentran para formar un ángulo.

array Objects displayed in rows and columns.

row → column

arreglo Objetos organizados en filas y columnas.

fila → columna

Bb

bar graph A graph that uses bars to show data.

gráfica de barras Gráfica que usa barras para ilustrar datos.

before

6 is just *before* 7

antes

6 está justo *antes* del 7

between

49 is *between* 48 and 50

entre

49 está *entre* 48 y 50

Cc

cent

I¢   I cent

centavo

I¢   I centavo

Cc

cent sign (¢) The sign used to show cents.

1¢   5¢

signo de centavo (¢) El signo que se usa para mostrar centavos.

1¢   5¢

centimeter (cm) A metric unit for measuring length.

centímetro (cm) Unidad métrica para medir la longitud.

circle A closed, round two-dimensional shape.

círculo Bidimensional redonda y cerrada.

compare Look at objects, shapes, or numbers and see how they are alike or different.

comparar Observar objetos, formas o números para saber en qué se parecen y en qué se diferencian.

cone A three-dimensional shape that narrows to a point from a circular face.

cono Figura tridimensional que se estrecha hasta un punto desde una base circular.

count back On a number line, start at the greater number (5) and count back (3).

contar hacia atrás En una recta numérica, comienza en un número mayor (5) y cuenta (3) hacia atrás.

$$5 - 3 = 2$$

$$5 - 3 = 2$$

count on On a number line, start at the first addend (4) and count on (2).

seguir contando En una recta numérica, comienza en el primer sumando (4) y cuenta (2) hacia delante.

$$4 + 2 = 6$$

$$4 + 2 = 6$$

Cc

cube A three-dimensional shape with 6 square faces.

cubo Figura tridimensional con 6 caras cuadradas.

cylinder A three-dimensional shape that is shaped like a can.

cilindro Figura tridimensional que tiena la forma de una lata.

Dd

data Numbers or symbols, sometimes collected from a survey or experiment, that show information. *Data* is plural.

Name	Number of Pets
Mary	3
James	I
Alonzo	4

datos Números o símbolos que se recopilan mediante una encuesta o experimento para mostrar información.

Nombre	Número de mascotas
María	3
James	I
Alonzo	4

day 1 day = 24 hours
Examples: Sunday, Monday, Tuesday, Wednesday, Thursday, Friday, Saturday

día 1 día = 24 horas
Ejemplos: domingo, lunes, martes, miércoles, jueves, viernes y sábado

difference The answer to a subtraction problem.

$$3 - 1 = 2$$
↑
The difference is 2.

diferencia Resultado de un problema de resta.

$$3 - 1 = 2$$
↑
La diferencia es 2.

digit A symbol used to write numbers. The ten digits are

0, 1, 2, 3, 4, 5, 6, 7, 8, 9.

dígito Símbolo que se utiliza para escribir números. Los diez dígitos son
0, 1, 2, 3, 4, 5, 6, 7, 8, 9.

digital clock A clock that uses only numbers to show time.

reloj digital Reloj que marca la hora solo con números.

dime dime = 10¢ or 10 cents

head · tail

moneda de 10¢ moneda de diez centavos = 10¢ o 10 centavos

cara · cruz

dollar one dollar =100¢ or 100 cents. It can also be written as $1.00 or $1.

front

back

dólar un dólar = 100¢ o 100 centavos. También se puede escribir $1.00.

frente

revés

dollar sign ($) The sign used to show dollars.

one dollar = $1 or $1.00

signo de dólar ($) Símbolo que se usa para mostrar dólares.

un dólar = $1 o $1.00

doubles (and near doubles) Two addends that are the same number.

$6 + 6 = 12$ ← doubles

$6 + 7 = 13$ ← near doubles

dobles (y casi dobles) Dos sumandos que son el mismo número.

$6 + 6 = 12$ ← dobles

$6 + 7 = 13$ ← casi dobles

edge The line segment where two *faces* of a three-dimensional shape meet.

edge →

arista Segmento de recta donde se encuentran dos caras de una figura tridimensional.

arista →

equal groups Each group has the same number of objects.

There are four equal groups of counters.

grupos iguales Cada grupo tiene el mismo número de objetos.

Hay cuatro grupos iguales de fichas.

equal parts Each part is the same size.

This sandwich is cut into 2 equal parts.

partes iguales Cada parte es del mismo tamaño.

Este sándwich está cortado en 2 partes iguales.

Ee

equal to =	igual a =

$6 = 6$	$6 = 6$
6 is *equal to* or the same as 6.	6 es *igual* o lo mismo que 6.

estimate Find a number close to an exact amount.	**estimar** Hallar un número cercano a la cantidad exacta.
47 + 22 rounds to 50 + 20. The estimate is 70.	47 + 22 se redondea a 50 + 20. La estimación es 70.

even number Numbers that end with 0, 2, 4, 6, 8.	**número par** Los números que terminan en 0, 2, 4, 6, 8.

expanded form The representation of a number as a sum that shows the value of each digit. Sometimes called *expanded notation*.	**forma desarrollada** La representación de un número como suma que muestra el valor de cada dígito. También se llama *notación desarrollada*.
536 is written as 500 + 30 + 6.	536 se escribe como 500 + 30 + 6.

half hour (or half past)
A unit to measure time.
Sometimes called *half past*
or *half past the hour*.

a half hour = 30 minutes

media hora (o y media)
Unidad para medir tiempo.
A veces se dice *hora y media*.

media hora = 30 minutos

halves Two equal parts of
a whole. Each part is a half
of the whole.

mitades Dos partes iguales
de un todo. Cada parte es
la mitad de un todo.

hexagon A 2-dimensional
shape that has six sides.

hexágono Figura
bidimensional que tiene seis
lados.

Hh

hour A unit to measure time.

1 hour = 60 minutes

hora Unidad para medir tiempo.

1 hora = 60 minutos

hour hand The hand on a clock that tells the hour. It is the shorter hand.

hour hand

manecilla horaria Manecilla del reloj que indica la hora. Es la más corta.

manecilla horaria

hundreds The numbers in the range of 100–999. It is the place value of a number.

365

3 is in the hundreds place.
6 is in the tens place.
5 is in the ones place.

centenas Los números en el rango de 100 a 999. Es el valor posicional de un número.

365

3 está en el lugar de las centenas.
6 está en el lugar de las decenas.
5 está en el lugar de las unidades.

inch (in) A customary unit for measuring length. The plural is *inches*.

12 inches = 1 foot

pulgada (pulg) Unidad usual para medir longitud.

12 pulgadas = 1 pie

inverse Operations that are opposite of each other.

Addition and subtraction are inverse or opposite operations.

operaciones inversas Operaciones que se oponen una a otra.

La suma y la resta son operaciones inversas u opuestas.

Kk

key Tells what (or how many) each symbol stands for.

clave Indica qué o cuánto representa cada símbolo.

Animal doméstico favorito			
Pez	☺	☺	☺
Perro	☺		
Gato	☺	☺	

Clave: ☺ = 2 votos

length How long or how far away something is.

length

longitud El largo de algo o lo lejos que está.

longitud

less than <

4 < 7

4 is less than 7.

menor que <

4 < 7

4 es menor que 7.

line plot A graph that shows how often a certain number occurs in data.

diagrama lineal Una gráfica que muestra con qué frecuencia ocurre cierto número en los datos.

measure To find the length, height, weight, capacity, or temperature using standard or nonstandard units.

medir Hallar la longitud, altura, peso, capacidad o temperatura usando unidades estándares o no estándares.

meter (m) A metric unit for measuring length.

I meter = 100 centimeters

metro (m) Unidad métrica para medir longitud.

I metro = 100 centímetros

minute (min) A unit to measure time. Each tick mark is one minute.

I minute = 60 seconds

minuto (min) Unidad para medir tiempo. Cada marca es un minuto.

I minuto = 60 segundos

minute hand The longer hand on a clock that tells the minutes.

minute hand

minutero La manecilla más larga del reloj. Indica los minutos.

minutero

Mm

missing addend The missing number in a number sentence that makes the number sentence true.

$$9 + \boxed{} = 16$$

The missing addend is 7.

sumando desconocido El número desconocido en un enunciado numérico que hace que este sea verdadero.

$$9 + \boxed{} = 16$$

El sumando desconocido es 7.

month A unit of time.
12 months = 1 year

April						
Sunday	Monday	Tuesday	Wednesday	Thursday	Friday	Saturday
		1	2	3	4	5
6	7	8	9	10	11	12
13	14	15	16	17	18	19
20	21	22	23	24	25	26
27	28	29	30			

This is the month of April.

mes Unidad de tiempo.
12 mesas = 1 año

Abril						
domingo	lunes	martes	miércoles	jueves	viernes	sábado
		1	2	3	4	5
6	7	8	9	10	11	12
13	14	15	16	17	18	19
20	21	22	23	24	25	26
27	28	29	30			

Este es el mes de abril.

Nn

near doubles Addition facts in which one addend is exactly 1 more or 1 less than the other addend.

casi dobles Operaciones de suma en las cuales un sumando es exactamente 1 más o 1 menos que el otro sumando.

nickel nickel = 5¢ or 5 cents

head   tail

moneda de 5¢ moneda de cinco centavos = 5¢ o 5 centavos

cara   cruz

number line A line with number labels.

recta numérica Recta con marcas de números.

Oo

o'clock At the beginning of the hour.

It is 7 o'clock.

en punto El momento en que comienza cada hora.

Son las 7 en punto.

odd number Numbers that end with 1, 3, 5, 7, 9.

números impares Los números que terminan en 1, 3, 5, 7, 9.

Oo

ones The numbers in the range of 0-9. A place value of a number.

65

5 is in the ones place.

unidades Los números en el rango de 0 a 9. Valor posicional de un número.

65

El 5 está en el lugar de las unidades.

order

1, 3, 6, 8, 10

These numbers are in order from least to greatest.

orden

1, 3, 6, 8, 10

Estos números están en orden de menor a mayor.

Pp

P.M. The hours from noon until midnight.

p.m. Las horas que van desde el mediodía hasta la medianoche.

parallelogram A two-dimensional shape that has four sides. Each pair of opposite sides is equal and parallel.

paralelogramo Figura bidimensional que tiene cuatro lados. Cada par de lados opuestos son iguales y paralelos.

partition To divide or "break up."

separar Dividir o desunir.

pattern An order that a set of objects or numbers follows over and over.

pattern unit

patrón Orden que sigue continuamente un conjunto de objetos o números insertar punto.

unidad de patrón

penny penny = 1¢ or 1 cent

head   tail

moneda de 1¢ moneda de un centavo = 1¢ o 1 centavo

cara   escudo

pentagon A polygon with five sides.

pentágono Polígono de cinco lados.

picture graph A graph that has different pictures to show information collected.

gráfica con imágenes Gráfica que tiene diferentes imágenes para ilustrar la información recopilada.

place value The value given to a *digit* by its place in a number.

1,365

1 is in the thousands place.
3 is in the hundreds place.
6 is in the tens place.
5 is in the ones place.

valor posicional El valor dado a un *dígito* según su posición en un número.

1,365

1 está en el lugar de los millares.
3 está en el lugar de las centenas.
6 está en el lugar de las decenas.
5 está en el lugar de las unidades.

pyramid A three-dimensional shape with a polygon as a base and other faces that are triangles.

pirámide Figura tridimensional con un polígono como base y otras caras que son triángulos.

Qq

quadrilateral A shape that has 4 sides and 4 angles.

cuadrilátero Figura con 4 lados y 4 ángulos.

quarter quarter = 25¢ or 25 cents

head   tail

moneda de 25¢ moneda de 25 centavos = 25¢ o 25 centavos

cara   cruz

quarter hour A quarter hour is 15 minutes. Sometimes called *quarter past* or *quarter til*.

cuarto de hora Un cuarto de hora es 15 minutos. A veces se dice hora y cuarto.

Rr

rectangle A plane shape with four sides and four corners.

rectángulo Figura plana con cuatro lados y cuatro esquinas.

rectangular prism
A three-dimensional shape with 6 faces that are rectangles.

prisma rectangular Figura tridimensional con 6 caras que son rectángulos.

regroup Take apart a number to write it in a new way.

reagrupar Separar un número para escribirlo de una nueva forma.

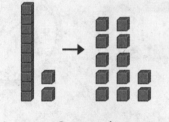

I ten + 2 ones becomes
12 ones.

I decena + 2 unidades se convierten en 12 unidades.

related fact(s) Basic facts using the same numbers. Sometimes called a *fact family*.

operaciones relacionadas Operaciones básicas en las que se usan los mismos números. También se llaman *familias de operaciones*.

$$4 + 1 = 5 \qquad 5 - 4 = 1$$
$$1 + 4 = 5 \qquad 5 - 1 = 4$$

repeated addition To use the same addend over and over.

suma repetida Usar el mismo sumando una y otra vez.

rhombus A shape with 4 sides of the same length.

rombo Paralelogramo con cuatro lados de la misma longitud.

round Change the *value* of a number to one that is easier to work with.

redondear Cambiar el *valor* de un número a uno con el que es más fácil trabajar.

24 rounded to the nearest ten is 20.

24 redondeado a la decena más cercana es 20.

Ss

side One of the line segments that make up a shape.

lado Uno de los segmentos de recta que componen una figura.

A pentagon has five sides.

El pentágono tiene cinco lados.

Ss

skip count Count objects in equal groups of two or more.

2, 4, 6, 8, 10

contar salteado Contar objetos en grupos iguales de dos o más.

2, 4, 6, 8, 10

---

slide To move a shape in any direction to another place.

slide

deslizar Traslador una figura a una nueva posición.

deslizar

---

sphere A three-dimensional shape that has the shape of a round ball.

esfera Figura tridimensional que tiene la forma de una pelota redonda.

square A two-dimensional shape that has four equal sides. Also a rectangle.

cuadrado Figura bidimensional que tiene cuatro lados iguales. También es un rectángulo.

subtract (subtraction) Take away, take apart, separate, or find the difference between two sets. The opposite of *add*.

$$5 - 5 = 0$$

restar (sustracción) Eliminar, quitar, separar o hallar la diferencia entre dos conjuntos. Lo opuesto de *sumar*.

$$5 - 5 = 0$$

sum The answer to an addition problem.

$$2 + 4 = 6$$

suma Resultado de la operación de sumar.

$$2 + 4 = 6$$

Ss

survey Collect data by asking people the same question.

Favorite Animal	
Dog	Ж|
Cat	Ж

This survey shows favorite animals.

encuesta Recopilar datos al hacer las mismas preguntas a un grupo de personas.

Animal favorito	
Perro	Ж|
Gato	Ж

Esta encuesta muestra los animales favoritos.

symbol A letter or figure that stands for something.

This symbol means to add.

+

símbolo Letra o figura que representa algo.

Este símbolo significa sumar.

+

Tt

tally marks A mark used to record data collected in a survey.

Ж|| ||

tally marks

marca de conteo Marca que se utiliza para registrar los datos recopilados en una encuesta.

ЖН ||

marcas de conteo

tens The numbers in the range of 10–99. It is the place value of a number.

65

6 is in the tens place.
5 is in the ones place.

decenas Los números en el rango de 10 a 99. Es el valor posicional de un número.

65

6 está en el lugar de las decenas.
5 está en el lugar de las unidades.

thirds Three equal parts.

tercios Tres partes iguales.

thousand(s) The numbers in the range of 1,000–9,999. It is the place value of a number.

1,365

1 is in the thousands place.
3 is in the hundreds place.
6 is in the tens place.
5 is in the ones place.

millar(es) Los números en el rango de 1,000 a 9,999. Es el valor posicional de un número.

1,365

1 está en el lugar de los millares.
3 está en el lugar de las centenas.
6 está en el lugar de las decenas.
5 está en el lugar de las unidades.

three-dimensional shape A shape that has length, width, and height.

figura tridimensional Que tiene tres dimensiones: largo, ancho y alto.

trapezoid A two-dimensional shape with four sides and only two opposite sides that are parallel.

trapecio Figura bidimensional de cuatro lados con solo dos lados opuestos que son paralelos.

triangle A three-dimensional shape with three sides and three angles.

triángulo Figura tridimensional con tres lados y tres ángulos.

two-dimensional shape The outline of a shape - such as a triangle, square, or rectangle - that has only *length* and *width*.

figura bidimensional Contorno de una figura, como un triángulo, un cuadrado o un rectángulo, que solo tiene *largo* y *ancho*.

vertex

vertex

vertex   vertex

vértice

vértice

vértice   vértice

week A part of a calendar.
1 week = 7 days

Sunday	Monday	Tuesday	Wednesday	Thursday	Friday	Saturday

semana Parte de un calendario
una semana = 7 días

domingo	lunes	martes	miércoles	jueves	viernes	sábado

whole The entire amount or object.

el todo La cantidad total o el objeto completo.

yard (yd) A customary unit for measuring length.

1 yard = 3 feet or 36 inches

yarda Unidad usual para medir la longitud.

1 yarda = 3 pies o 36 pulgadas

year A way to count how much time has passed or will pass. 1 year = 12 months

año Un período insertar punto, un año = 12 meses

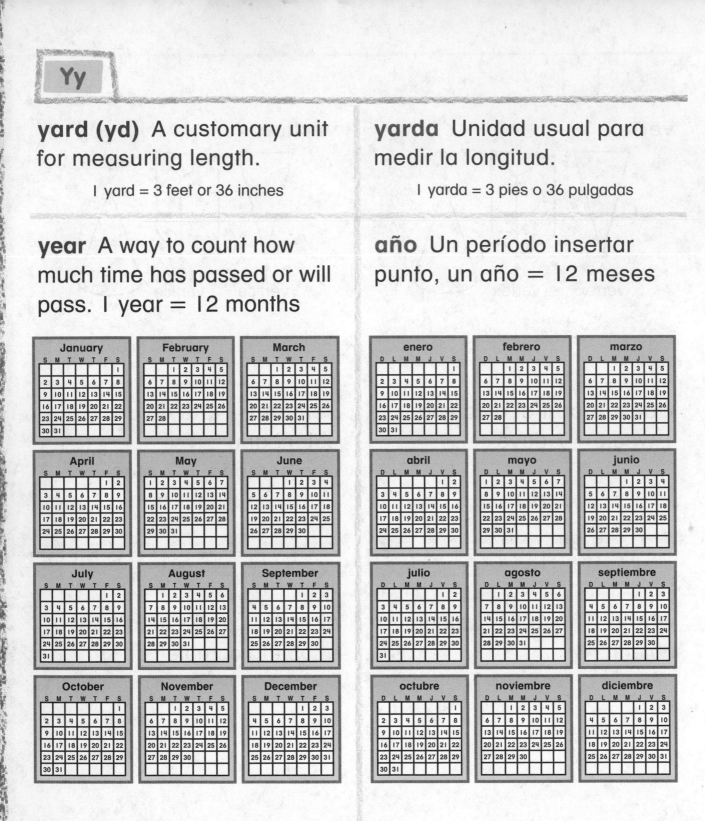

Name _____

Work Mat 3: Number Lines

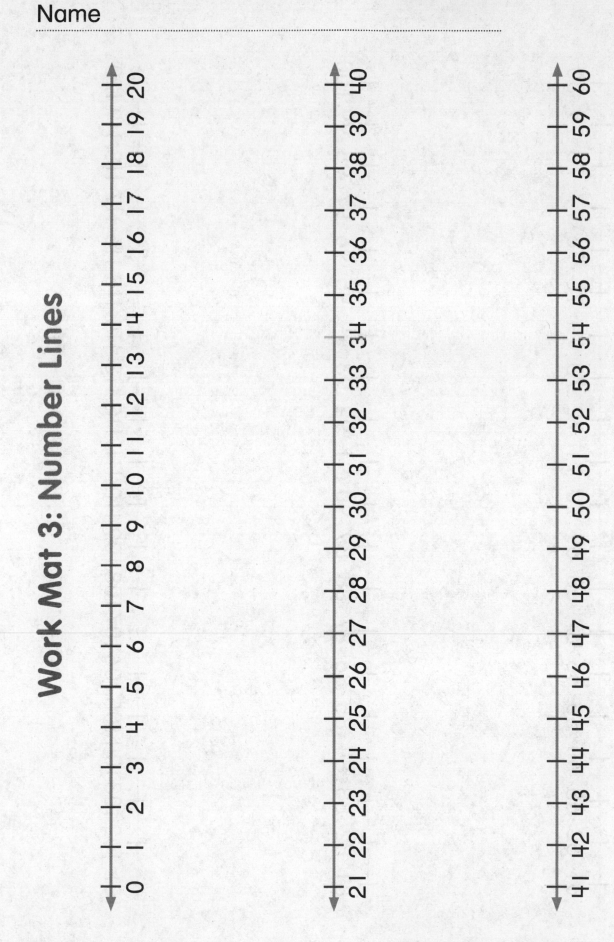

0 1 2 3 4 5 6 7 8 9 10 11 12 13 14 15 16 17 18 19 20

21 22 23 24 25 26 27 28 29 30 31 32 33 34 35 36 37 38 39 40

41 42 43 44 45 46 47 48 49 50 51 52 53 54 55 56 57 58 59 60

Work Mat 4: Number Lines

61 62 63 64 65 66 67 68 69 70 71 72 73 74 75 76 77 78 79 80

81 82 83 84 85 86 87 88 89 90 91 92 93 94 95 96 97 98 99 100

101 102 103 104 105 106 107 108 109 110 111 112 113 114 115 116 117 118 119 120

Work Mat 4: Number Lines